101 Cool Smartphone Techniques

101 Cool Smartphone Techniques:

Covers Series 60® Phones from Nokia®, Samsung®, Siemens™, Panasonic®, Sendo®, and More!

Dean Andrews

Wiley Publishing, Inc.

101 Cool Smartphone Techniques

Published by
Wiley Publishing, Inc.
10475 Crosspoint Boulevard
Indianapolis, IN 46256
www.wiley.com

Library of Congress Card Number: 2004118280

ISBN: 0-7645-7942-8

Manufactured in the United States of America

10 9 8 7 6 5 4 3 2 1

1B/QS/QS/QV/IN

For general information on our other products and services or to obtain technical support, please contact our Customer Care Department within the U.S. at (800) 762-2974, outside the U.S. at (317) 572-3993 or fax (317) 572-4002.

Wiley also publishes its books in a variety of electronic formats. Some content that appears in print may not be available in electronic books.

About the Author

Photo: Jane Andrews

Dean Andrews works in research and development at Nokia. Every day he discovers something new you can do with cellular phones. Dean has contributed to several computer books and wrote *Windows 98 Hints & Hacks* (Que), which was a finalist in the Best Advanced How-To Book category of the Computer Press Association Book Awards of 1998. He has written articles for *PC World* magazine, *Macworld* magazine, CNET, Boston.com, and other magazines and Web sites. Mr. Andrews and his family live in Boston.

Credits

Senior Acquisitions Editor
Mike Roney

Project Editor
Elizabeth Kuball

Technical Editor
Brian Woods

Copy Editor
Elizabeth Kuball

Editorial Manager
Robyn Siesky

Vice President and Executive Group Publisher
Richard Swadley

Vice President and Publisher
Barry Pruett

Project Coordinator
April Farling

Graphics and Production Specialists
Kelly Emkow
Lauren Goddarg
Jennifer Heleine
Heather Ryan

Quality Control Technicians
Leeann Harney
Jessica Kramer
Brian H. Walls

Proofreading and Indexing
TECHBOOKS Production Services

To Jane — with friendship, love, and loyalty

Acknowledgments

This book would not be in print if it weren't for Michael Roney, a senior acquisitions editor at Wiley Publishing, Inc., and David Fugate, my agent at Waterside Productions, Inc. These two listened to me ranting about a smartphone tips book long before anyone else. To get this book project started, the three of us had to work hard for several months. I thank both of them for their tireless dedication.

Inside Nokia, Brian Woods, Ari Hakkarainen, Esa Eerola, and Dan Shugrue will forever be my heroes. These gentlemen also listened to my schemes for an end-user book about cellphones and took action. They garnered Nokia's support for this project in meeting after meeting. Brian Woods even signed on as Technical Editor. I thank them all. Also, inside Nokia, I must thank Laurie Armstrong, David Dzumba, Demetrios Boutsikakis, Bridget Winston, and Carol Perletz.

My editor, Elizabeth Kuball, and the entire staff at Wiley are talented and dedicated. I'm proud that such a stellar crew published this book.

My kids — Luke, Veronica, Declan, and an upcoming as yet unnamed arrival — pulled me away from my computer, demanded attention, energy, time and daily "shaky" hugs. Together, they kept my spirit alive and reminded me the role of dad is far more important than the role of author. I thank them for that.

Living in different states, I don't talk with my mother, father, and my brother Glenn nearly enough. But I love them, and their support keeps me moving forward each day. Two good friends, Steve Reiling and Marcus Torchia, also deserve thanks for keeping me laughing and typing, sometimes in the middle of the night.

The entire, extended Davey family has always been there to help with anything at any time. It does indeed take a "village" and a large, loving clan to raise children, to get by each day, and to write a book. I thank them all.

Most importantly, my wife, Jane, inexplicably put up with me having a full-time job and writing this book during nights and weekends. Her energy, support, and love humble me. She makes my dreams come true.

Getting this book project off the ground took more than ten months of hard work — pitching the book to publishers and to phone makers, getting rejected, refining the idea, and starting over again. Was it all worth it? Yes! The experience of creating this book has confirmed for me one of life's most important lessons — pursue your dreams relentlessly. Never give up!

— D.A.

Contents at a Glance

● ●

Contents

• •

Introduction

Welcome!

It doesn't take a genius to know that if you're reading this book, you very likely own a new or new-ish smartphone. Congratulations! You have a very cool gadget in your hands.

The thing is, it's cooler than you realize. And, that's what this book is all about. In these pages, I show you how to do more with your phone than you knew it could do.

Sure, you know how to make and take phone calls. You know your phone features a color screen. You know how to snap some pictures with your phone's camera. You may even know how to send and receive picture — or multimedia — messages.

But that, my friend, is really just the beginning.

A powerful convergence of events — advances in phone operating systems, hardware-integration techniques, new processor chips, cheaper memory, new technologies like Bluetooth, and strong competition among cellphone makers — has created a new category of devices, namely, the smartphone, also known as the camera phone.

For several reasons — including phone makers not wanting to overwhelm people with too much information — phone user manuals only scratch the surface of what these devices can really do. This book shows you the next level of tips and techniques for taking advantage of these powerful new cellphones.

Also, user manuals, by their very nature, only show you how to use a phone. This book shows you how to *make use* of your phone to make you more productive, bring more fun into your life, help you customize your phone to suit your lifestyle, cause you fewer headaches, and harmonize your information and data easily across all your computers and devices.

I worked hard to choose the very best and coolest techniques. Most of them you won't find documented anywhere else — including user manuals, techie Web sites, magazines, or books. I wrote this book with you in mind. I've tried hard to make it fun, interesting, and most of all useful!

Get your phone out and put it next to you as browse through the techniques. Try a few. You'll see that this book is worth every penny you spent on it.

Read on and enjoy!

Which phones are covered in this book

You may have noticed the subtitle of the book: "Covers Series 60 phones from Nokia, Samsung, Siemens, Panasonic, Sendo, and more!" What the heck is a Series 60 phone? Glad you asked.

The Series 60 platform is a software package that many mobile-phone manufacturers have licensed and use to help them build mobile handsets with advanced data capability that are typically optimized for use with one hand. The common label for such one-hand-operated mobile phones is *smartphone*. Manufacturers that have licensed this software and use it in some of their models include Lenovo, LG, Nokia, Panasonic, Samsung, Sendo, and Siemens. The Series 60 platform consists of two parts: a suite of application enablers, which lets you send text messages, send e-mail, and browse the Internet, for example; and a user interface, which basically defines and controls how you interact with the phone. The Series 60 user interface is really the key to this book. The techniques work on a variety of phone makes and models, because even though the phones look different on the outside, the phones all feature the same user interface and core application enablers under the covers. I explain a little more about the Series 60 platform and its underlying Symbian operating system in the "Anatomy of a smartphone" section toward the end of this Introduction.

Don't feel dumb for not knowing what Series 60 or Symbian means. Very few people outside of the wireless industry have ever heard the terms. However, the number of available Series 60–based phones is rapidly growing, and the phone makers that use Series 60 software plan to spread the word about this common platform in the coming months. But, by reading this book, you're getting clued in before most.

First and most importantly, I'm not trying to fool anyone. The tips in this book don't work for every smartphone or camera phone on the planet. However, the tips *do* work for some of the best-selling smartphones in the world, from a variety of manufacturers.

The goal of this section is to help you determine if your particular phone is covered or not covered. It's a quick two-step process:

Step 1: Find your phone on the following list. Do you know your phone's make and model or can you match your phone to one of the following pictures? The following phone makes and models are definitely covered in these pages:

Sendo X

Siemens SX1

Panasonic X700

Nokia 3620/3660

Nokia 3650

Nokia 6260

Nokia 6600

Nokia 6620

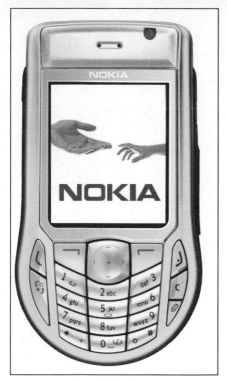

Nokia 6630

Nokia 7610

Nokia 7650

Nokia N-Gage
Game Deck

Nokia N-Gage
QD Game Deck*

Samsung SGH-D710

Note: The Nokia N-Gage and Nokia N-Gage QD Game Decks do not have integrated digital cameras. However, both are Series 60–based phones, and every technique in this book except for those specifically requiring you to capture an image with the phone work with them. Keep in mind that you can easily send or transfer images to these game-deck phones and still perform the tips that involve using an image.

If your phone is in this list, the techniques in this book will work on your phone. Not sure of your phone's make and model? Simply remove the back cover and the battery, and underneath you'll see a white label that, among other things, includes information such as the manufacturer name and the model number of the handset. Alternatively, you can check your user manual or ask your service provider about the make and model of your phone.

Don't see your phone on the list? Go to Step 2.

Step 2: Visit the Series 60 Website (www.series60.com). This is the official Web site of the Series 60 platform. This site gets frequently updated with the latest Series 60–based phones.

You'll see a picture of every Series 60–based handset as well as the makes and models. If you see your phone on this Web site, the techniques in this book will work for you.

Tada! You're done. Easy, wasn't it?

After following these two steps, you should know whether your phone can perform the techniques described in this book.

If you find your phone is *not* covered by the book. I want you to:

+ **Put this book back on the shelf where you found it.** If you're in a bookstore, go browse the new-fiction shelves, pick out something nice for yourself, and go buy it with the money you were going to spend on this book. Go ahead—you deserve it!

+ **Then send me an e-mail at** dean@deanandrewsmobile.com **and ask me to write a new tips book with your phone in it!**

+ **Finally, point your browser to** www.deanandrewsmobile.com **where I'll be posting more phone tips for more phone makes and models.**

If, on the other hand, your phone is covered by this book. I want you to:

+ **Hold the book tightly with both hands.** Others around you in the store might try to snatch it away from you to get at these awesome techniques. Raise your elbows like a basketball player to avoid being stripped of the book by an aggressive bookworm. If only one or two copies of the book remain on the shelf you might be in real danger, so walk briskly toward the store's checkout counter and buy the book as quickly as possible.

+ **Then send me an email at** dean@deanandrewsmobile.com **and let me know what you think of the book.** You'll also find more phone tips for your phone at www.deanandrewsmobile.com.

Who should read this book

You!

All kidding aside, some tips books target an audience of only complete novices, or only business users, or, in a few cases, only tech-savvy gurus. Not this book.

This book's 101 techniques cover a broad spectrum of purposes. Many different kinds of people will find them useful, including teenagers, businesspeople, parents, students, seniors, techie people of all ages, and others.

The step-by-step instructions are written so that someone who has never used a computer or smartphone before will be able to perform the techniques. But I include enough tips and extra information that even a technical guru will learn something new.

These new phones aren't cheap. You, or someone who gave you the phone as a gift, spent a few bucks on one. Unfortunately, surveys show that many people only use a small percentage of the incredible features of the new generation of cellular phones.

That's why I wrote this book. You'll find that Series 60–based phones offer tremendous value for the price, and I want to ensure that you get the biggest bang for your buck. You'll find out how in these pages.

Terms you need to know

The wireless world brings a lot of new terminology and acronyms with it. Many of them you don't really need in your day-to-day life. Learning a select few, though, will help you make better use of your phone and help you quickly grasp some of the techniques in this book. Don't worry. I kept the list short and I won't bore you with unnecessary details.

So, scan through this list. At the very least, you'll be able to throw out a few cool buzzwords at your next social function.

+ **SMS (Short Message Service):** Also known as *text messaging,* these text messages — up to a maximum of 4,800 characters in length (but sometimes limited in practice to shorter lengths by your network) — are sent to and from phones through a service-provider network. SMS is the first major and still most popular cellphone feature after voice calling. The most common usage of SMS is sending messages from one phone to another, but you can also send SMS messages from some Web sites to phones. There are also special types of SMS messages that can configure your phone settings automatically. I discuss these in Chapter 1.

  Check you service plan for how many SMS messages you can send per month and how much it costs you if you go over your limit.

+ **MMS (Multimedia Message Service):** Also, known as *picture messaging* and *video messaging,* this service marked the evolution of SMS, to allow users to essentially send text messages with audio, image, and video attachments from phone to phone or phone to e-mail. Although many wireless-industry players hope that MMS traffic will grow to SMS-traffic levels, adoption of multimedia messaging thus far has been slower than expected.

  MMS traffic may be treated differently than SMS by your service provider, so check your plan to determine how many MMS messages you can send per month and how much it costs you if you go over your limit.

✦ **GSM (Global System for Mobile communications):** GSM is a wireless network technology most popular in Europe but growing in popularity within the United States and other countries. There are currently over 1 billion GSM cellphone subscribers in over 200 countries. GSM supports international roaming. There are competitive wireless networking systems like TDMA and CDMA. The next evolution of wireless networks is 3G (third generation), and the two main standards are WCDMA (based on GSM) and CDMA2000 (based on CDMA). The Series 60–based Nokia 6630 handset is an example of a handset with WCDMA capability.

The keys things you should know as an end user are that every current Series 60–based phone uses GSM-based technology (one even uses WCDMA) and that every GSM phone uses a SIM card. For more information, browse www.gsmworld.com.

✦ **SIM card (Subscriber Identity Module card):** This small printed circuit board is generally about the size of a postage stamp. When inserted into a phone, a SIM card gives a phone a unique identity; the wireless network uses it to map calls to a particular phone, track phone usage, process billing, and so on.

The SIM card also works as a storage location and holds a phonebook and special security information. You may never have seen your SIM card. It generally sits underneath the phone's battery inside the phone's cover and is often inserted by the service provider before you receive your phone. I discuss SIM cards in some of the techniques in this book.

✦ **3GP:** This is a video format particular to cellphones. In the very earliest days of camera phones, you had to convert a 3GP-formatted file to a PC video format like QTM (QuickTime) or AVI (Windows Media Player) before you could see the video on your PC. But now, PC video player applications are adding 3GP to the formats they support. So, in some cases, you can just transfer phone-recorded video to your PC and play it there. I discuss this more in Chapter 7. For additional information and access to 3GP PC media players, browse www.3gp.com.

✦ **Bluetooth:** This is a personal wireless network technology that comes embedded in many of today's mobile devices (mobile phones, phone accessories like headsets and keyboards, PCs, automobiles, and so on). The unusual name comes from ambitious Viking in the tenth century — Danish King Harald Bluetooth — because he united Denmark and Norway. Get it? United? Networking? Neither do I. Anyway, this technology enables your phone to wirelessly connect to and share data with other types of devices, like computers, automobiles, and other mobile phones. You find out exactly how in this book. For more information from the Bluetooth SIG (Special Interest Group), point your browser to www.bluetooth.com.

✦ **Infrared:** This is the wireless light-based technology that makes your TV remote control work. Although your remote is a one-way (remote to TV or VCR or DVD player or TiVo) communication path, infrared (IrDA) technology also offers two-way communication. That's why it comes embedded in many mobile phones. Like Bluetooth, you can connect your phone to PCs, PDAs, and other phones using IrDA. For more information, and a not-too-technical description of exactly how IrDA works, point your browser to www.irda.org.

✦ **Memory cards:** Like SIM cards, memory cards are about the size of a postage stamp. These flash memory cards come in varying sizes formats, like MMC (Mulitmedia Memory Card) or SD (Secure Digital) and support memory ranges from 16MB up to over 512MB. You can expand your phone's memory—useful for storing digital pictures, video, MP3 tunes, and more—by inserting a memory card. You'll learn more about memory cards and how to put them to use in this book.

✦ **Standby mode:** The state of your phone when it's powered up but you're not using it. In other words, it's standing by, ready to receive calls. For Series 60–based phones in standby mode, you'll typically see a screen showing your service provider's name at the top. Some of the techniques described in this book start from the standby mode screen.

✦ **Service provider/mobile operator:** In the context of this book, it's the carrier (AT&T Wireless, Cingular, Orange, T-Mobile, Verizon Wireless, Vodafone, and so on). These are the companies that provide the service plan for your phone. They do not, generally, manufacturer cellphones; they're the main channel by which phones and service plans are sold to the consumer. For performing these techniques, all you have to be able to do is know the name of your service provider, because you'll need to access its Web site or choose its name from a list in some of the step-by-step instructions.

Anatomy of a smartphone

At some point in the not-too-distant future, cellphones that only make and receive calls will go the way of vinyl records, rotary dial phones, and cigarette smoking in public places. Because we're still in the early days of smartphones, I'm guessing you're still very familiar with a basic single-function cellphone— monochrome screen, a couple of menu keys, a phonebook, and maybe a way to change the very simplistic ring tones.

In terms of "intelligence," these old cellphones weren't much smarter than your TV remote control and they were a lot less smart than, say, TiVo. These basic phones had no ports to communicate with any other devices. They had only one-note-at-a-time monophonic ring tones. They did not support synchronizing with PCs, couldn't run Java applications, and so on.

But, times have changed. Dramatically.

In the following sections, I break down the smartphone into four basic elements: hardware, operating system, user interface, and add-on applications.

Hardware

Hardware specifics of Series 60–based phones vary slightly from manufacturer to manufacturer, but your phone likely features most or all of the following:

✦ **Keys and display:** Although Series 60 phones come in many shapes and sizes, they have the same basic key layout (see the figure). Please take a moment to look in particular at the key names — left and right menu keys, main menu key, send and end keys, joystick, and so forth. I use these names in the step-by-step instructions in this book. The LCD screens also vary slightly from phone to phone, but in general, the screens are 1½ x 2 inches of displayable area that can show up to 64,000 colors.

General Series 60 UI Hardware

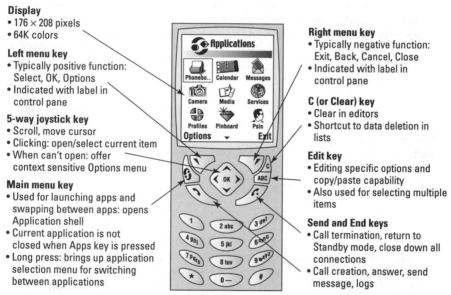

Display
• 176 × 208 pixels
• 64K colors

Left menu key
• Typically positive function: Select, OK, Options
• Indicated with label in control pane

5-way joystick key
• Scroll, move cursor
• Clicking: open/select current item
• When can't open: offer context sensitive Options menu

Main menu key
• Used for launching apps and swapping between apps: opens Application shell
• Current application is not closed when Apps key is pressed
• Long press: brings up application selection menu for switching between applications

Right menu key
• Typically negative function: Exit, Back, Cancel, Close
• Indicated with label in control pane

C (or Clear) key
• Clear in editors
• Shortcut to data deletion in lists

Edit key
• Editing specific options and copy/paste capability
• Also used for selecting multiple items

Send and End keys
• Call termination, return to Standby mode, close down all connections
• Call creation, answer, send message, logs

Learning the names of the keys will help you perform the techniques in this book.

✦ **Memory:** Generally, Series 60 phones feature 6MB to 10MB of integrated flash memory and also accept external memory cards for memory expansion (see "Terms you need to know" for more information).

✦ **Integrated digital camera:** Most digital cameras in mobile phones use CIF or VGA technology, yet some of the newer models feature resolutions up to 1.3 megapixels. In the very near future, with newer phones, pixel resolution on mobile phones will increase to several megapixels, much like what you'd find on traditional digital cameras.

✦ **Bluetooth:** A personal network technology (see the "Terms you need to know" section). Techniques in this book show you how to connect to other phones, PDAs, and PCs using the integrated Bluetooth capability.

✦ **Infrared:** Another mechanism for communication between your phone and other phones, devices, and PCs (see the "Terms you need to know" section).

Operating system

That's right, your phone has an operating system (OS) just like your computer or PDA. All Series 60–based phones feature an operating system built by Symbian Ltd. Symbian is an operating system designed for mobile phones and used by some of the biggest names in cellular. Symbian began operating as an independent company started by Motorola, Ericsson, Nokia, and Psion in 1998, with a goal of producing a phone operating system that could be commonly used by several cellphone manufacturers. Now, there over 18 Symbian-based phones available and many more soon to ship. Since it's launch, the ownership of the company has evolved. Ericsson, Nokia, Panasonic, Samsung, Siemens, and Sony Ericsson now jointly own Symbian. The Symbian OS is licensed externally to manufacturers, and licensees include the owners as well as other manufacturers, like LG and Motorola.

Series 60 is not the only software platform that can run on top of the Symbian operating system. Nokia offers a few other Symbian-ready platforms, including Series 40, Series 80, and Series 90. However, Series 60 is the only one that it licenses to other phone makers. Another licensable software platform for the Symbian OS is designed for pen-based mobile handsets and is called UIQ. UIQ is licensed by a company called UIQ Technologies.

Important features that Symbian delivers include the following:

✦ **Multitasking between applications:** This is the ability to pause one application and run another, and another, and then jump back to one of the other applications and pick up where you left off. Very few other (if any) mobile phone operating systems offer multitasking capabilities that are this extensive.

✦ **Running both native Symbian applications (C++) and Java (MIDP 1.0 and MIDP 2.0) applications**

✦ **Support of international languages**

✦ **Data synchronization (SyncML)**

✦ **Support of messaging protocols (SMS, MMS, and EMS) and Internet e-mail (POP3, IMAP4, and SMTP) standards**

✦ **Wide area networking protocols like TCP/IP, Ipv4, Ipv6, and WAP, as well as personal networking technologies like infrared, Bluetooth, and USB**

For more information on Symbian, point your browser to the official Symbian Web site (www.symbian.com).

User interface

The Series 60 platform is a suite of application enablers and a user interface that is optimized for the Symbian OS. Series 60 was built and is owned by Nokia. Nokia uses the Series 60 platform is many of its own phones and also licenses the software as an OEM (original equipment manufacturer) product to many other major phone makers. In fact, current licensees include leading handset manufacturers such as LG, Legend, Nokia, Panasonic, Samsung, Sendo, and Siemens.

Series 60 offers both a powerful customizable user interface and a suite of integrated application enablers. Here are some of the important features that Series 60 delivers:

+ Text and multimedia messaging

+ PIM (personal information manager) applications like Contacts and Calendar

+ Media Gallery organizer for your images, video, and sound files

+ Internet browsing

+ Digital still-image and video capturing

+ Connectivity to PCs and other devices

+ Video and audio streaming

+ E-mail

+ 3-D graphics support for games

Currently there are two "editions" of the Series 60 platform available on phones. The Nokia 3620, 3650, 3660, 7650, N-Gage Game Deck, N-Gage QD Game Deck, Siemens SX1 and the Sendo X all use Series 60 1st Edition, while all other phones (as of this writing) use Series 60 2nd Edition. There are some differences in terms of which application enablers are available in each edition. I point out these differences as appropriate throughout this book. Bottom line: If your phone uses Series 60 1st Edition, there will be a handful of techniques in this book that you won't be able to perform. I describe alternatives to reach the same goal or perform the same task within the techniques.

For more information about Series 60, point your browser to the official Web site (www.series60.com).

Add-on applications

Your phone can run native Symbian applications (C++) and Java applications (MIDP 1.0 and MIDP 2.0). I explain where to find cool applications and how to download and install them in the pages that follow.

How to use this book

First, find out if the phone you own is a Series 60 phone by reading the "Which phones are covered in this book" section, earlier in this Introduction.

Second, scan through the "Terms you need to know" and "Anatomy of a smartphone" sections, also in this Introduction, to understand some important terminology and the basic key names of your phone.

Finally, flip through the table of contents and pick a technique that looks interesting to you. Or, if you're daring, flip the pages and stop at one at random. I bet, using either method, you'll find a technique that will wow you!

From there, it's up to you. Please don't feel you need to read the book straight through like a novel. Feel free to jump from chapter to chapter, even technique to technique, following concepts — like productivity enhancement, customization, using multimedia, and so on — that compel you.

While you're reading, you'll come across the following icons used in this book:

 Tip   A tip describes a little known fact. It can also show another way to perform the same action currently being illustrated in the technique you're reading.

 Caution   A caution alerts you to information you need to be aware of.

 Note   A note describes general information about the current topic.

A final note about how to use this book: The Series 60 platform allows phone makers the ability to customize icons and menus and move objects around in different folders. This means a new Series 60–based phone might feature slightly different names for folders and menus like Gallery, Messaging, Tools, and so forth. The techniques in this book will still work for your phone — you just need to apply a little common sense about the folder and icon names you find on your smartphone.

Buying and Setting Up Your New Smartphone

I know — sometimes the cellphone ads come so fast and furious that you feel like you're under attack. Service providers try to outdo each other every week or so by offering more strangely categorized minutes, or new and different phones, or wacky services — or all three mixed together.

So, how can you cut through the confusion? If you just want a phone, how do you determine which is the best deal? And how do you know whether you've got the right service plan to go with it?

In this chapter, I arm you with shopping techniques and tips on setting up a new cellphone. You find out how to shop for the best deal on a phone before you decide on a plan. And, if you already want a particular phone, I show you how to buy only a phone over the Web and find a plan later.

After you have a new phone and a plan, this chapter shows you how to easily configure the phone for sending e-mail, browsing the Web, and sending multimedia (MMS) messages, as well as how to move your old phone's data to your new phone so you won't need to reenter all your contacts and other information from scratch.

Finally, make sure you register your phone with your service provider and check in with your phone's manufacturer. The last techniques in this chapter show you the goodies you'll often find at many service provider's Web sites.

Technique 1: Shopping for Deals over the Web

Don't ignore the barrage of print and newspaper ads for smartphones and service plans. They alert you to the latest ways in which service providers are competing with one another — more anytime minutes, cooler phones, or interesting combination deals with multiple phones and shared minutes.

If you have the luxury of time, check at least one or two Web sites in these three basic categories:

✦ Comparison-shopping Web sites

✦ Phone-manufacturer Web sites

✦ Service-provider Web sites

When you've investigated all three, you can be sure you're getting the best possible deal on your phone and service plan. The following sections provide a little more information about each category.

Comparison-shopping Web sites

Comparison-shopping sites search the Web for deals on both phones and service plans. They let you pick a particular phone and then show you all the different prices and matching service plans being offered all over the Web. You can very quickly determine the average price of a particular phone. The same goes for what you would end up paying in terms of a monthly service fee (see Figure 1-1). Some of the best sites to check include the following:

✦ **MyRatePlan.com:** www.myrateplan.com

✦ **Point.com:** www.point.com

✦ **TeleBright:** www.telebright.com

✦ **Wireless Guide:** www.wirelessguide.org

Need to know if a phone offers Bluetooth and infrared connections? Sites like Point.com and WirelessGuide.org allow you to quickly and easily compare features from phone to phone (see Figure 1-2).

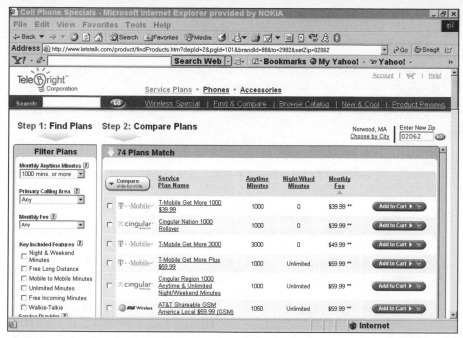

Figure 1-1: Sites like TeleBright show you the price ranges for phones and the service plans that go with them.

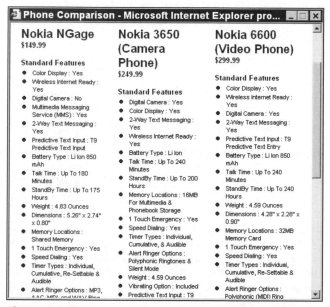

Figure 1-2: Comparing phone features is a breeze on sites like Point.com.

Phone-manufacturer Web sites

Fallen in love with a particular phone manufacturer's phones? Not a problem. Browsing the phone maker's Web site is a good idea. Not only will you get a full run-down on phone features, but in most cases you can buy phones and services plans directly from phone-maker sites.

Why would you ever shop for a phone without looking for a service plan? In some cases, that's the only way you're going to find the latest, greatest, and coolest Series 60 phone. For example, while I was writing this book, I purchased a Siemens SX1 over the Web without a service plan. The phone was commercially available in Europe but not from any U.S.-based service provider. So, I searched and found an SX1 through eBay.

Cellphones can be locked by software in such a way that they only work on a particular service provider's network, even though another carrier may support the same frequency bands. If you buy a locked phone without a service plan, chances are you'll never be able to get it working. When you buy a phone without a plan, make sure the phone you buy is *unlocked.* (If you're buying a phone with a service plan, don't worry about whether it's locked or unlocked — your service provider will make sure the phone works properly for you.)

The popular auction site eBay (www.ebay.com) has become a real source for phones. In addition to bidding in an auction for a phone, many eBay phone sellers allow you to buy a phone for a fixed price without bidding — more like a regular retail purchase over the Internet. That's how I purchased the SX1 I mention earlier — I used the "Buy It Now" offer from a phone seller, as opposed to buying through a regular auction.

If you do use eBay, make sure you check the feedback rating of the seller (eBay's basic measure of buyers' and sellers' trustworthiness) and read the feedback comments to determine if other buyers had any problems with the products sold by the particular seller offering the phone you want. *Remember:* Make sure the eBay seller says the phone is unlocked.

If you read the Introduction of this book, you know that all Series 60 phone are GSM phones. And you also know that all GSM phones use SIM cards. (SIM stands for *Subscriber Identity Module.*) SIM cards hold all relevant user information. SIM cards are the secret to being able to use phones that you purchase without a service plan. If you currently have a GSM phone and an active service plan, you can get started with your new unlocked GSM phone, simply by inserting your SIM card into it. Read the user manual of your current GSM phone for instructions on how to remove your SIM card (the SIM card is usually underneath the battery behind the back cover), and read the user manual of your new unlocked GSM phone to find out how to insert a SIM card into it.

One final note about buying phones without a service plan: You'll likely be paying full price for these phones. In the current business model of cellphone service providers, the service provider deeply subsidizes the cellphone itself—offsetting this subsidy by holding you to at least a one-year service-plan contract. So only choose the unlocked phone option if you can afford it. If you're on a budget, go with one of the discounted offers from the service provider, where the phone is either free or within your budget.

Most of the Series 60 phone-maker Web sites offer tools that help you choose among the many phone models these cellphone giants produce (see Figure 1-3). Additionally, you can quickly view all the accessories—like a car kit, a headset, and a desk stand (see Chapter 17 for more information). Here's a list of the Series 60 phone-manufacturer Web sites:

+ **LG:** www.lge.com

+ **Lenovo:** www.lenovo.com

+ **Nokia:** www.nokia.com

+ **Panasonic:** www.panasonic.com

+ **Samsung:** www.samsung.com

+ **Sendo:** www.sendo.com

+ **Siemens:** www.siemens-mobile.com

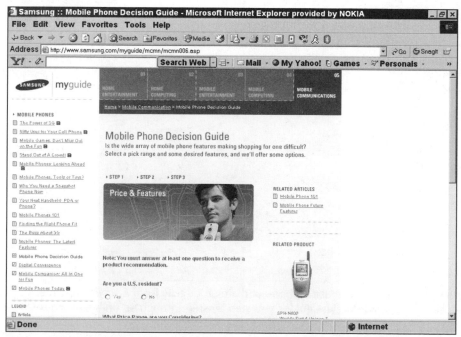

Figure 1-3: Series 60 phone-manufacturer Web sites offer tools to help you choose a phone. Here, Samsung's Web site asks you questions and then shows you its recommendations.

Service-provider Web sites

Although there are many service providers around the world, probably only a handful service the area where you live and work. Spend a little time investigating the offers on the Web sites of all the service providers in your area. Not only can you browse through their offerings, but you can also buy a phone and a plan online (see Figure 1-4). Here's a short list of some of the major cellphone service providers in the United States:

+ **AT&T Wireless:** www.attwireless.com

+ **Cingular:** www.cingular.com

+ **Orange:** www.orange.com

+ **T-Mobile:** www.t-mobile.com

+ **Verizon Wireless:** www.verizonwireless.com

+ **Vodafone:** www.vodafone.com

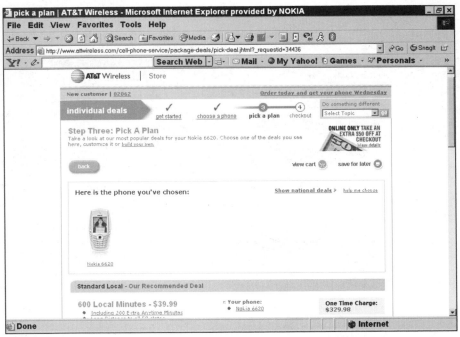

Figure 1-4: You can purchase a phone and a plan directly from a service provider's Web site. Here, AT&T Wireless guides you step by step through an online purchase.

Technique 2: Requesting Special-Configuration SMS Messages from Your Carrier and Manufacturer

When users purchase phones with advanced functionality, like browsing and MMS messaging, they typically also have to sign up for an appropriate plan with their service provider. These phones sometimes come preconfigured from the service providers, in terms of setup for these and other advanced features. This preconfiguration makes it easy to start using your phone for sending messages and such right after you get it. Unfortunately, not all service providers preconfigure all their phones, so there is a chance you'll receive a phone that hasn't been set up in advance.

Note   If you buy a phone without a plan over the Web as described in Technique 1, you'll need to use this technique to configure your new phone.

Navigating through your phone's settings menus can be intimidating. Choose Settings ➪ Connection ➪ Access Points, and you'll see items like Connection Name, Data Bearer, and Access Point Name (see Figure 1-5). Even if you have a technical background and are familiar with these terms, you still need to call your service provider's technical-support personnel (and perhaps spend some time queuing behind other information seekers) in order to learn the specific entries that will make your phone able to send and receive MMS messages, browse the Web, and so on.

Determining if your phone is preconfigured

How do know if the phone you've purchased is already preconfigured and doesn't need to be configured via special SMS messages?

✦ **Try sending a multimedia (MMS) message to a friend, colleague, or even yourself.** If the message successfully gets through, you know your phone is already configured properly for an MMS access point. Also, have someone send you an MMS to confirm MMS reception settings. Next try browsing the Web and sending an e-mail. If all these actions work, you don't need to configure your phone with SMS configuration messages.

✦ **Ask your service provider's sales staff or technical-support representatives if your phone comes preconfigured for MMS, e-mail, and Web browsing.** A great time to ask is when you receive your phone.

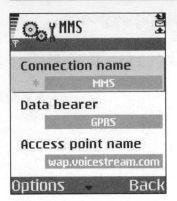

Figure 1-5: Trying to configure your own MMS and Web browsing phone settings can be intimidating.

Fortunately, you don't have to understand these terms, and you don't have to call anyone for help in order to configure these settings properly. Your phone manufacturer and service provider have already taken the guesswork (and the hard work) out of configuring phones by leveraging a special SMS-based technology that lets these companies send special-configuration SMS messages rather than just text messages.

 Note These special-configuration messages are also called OTA (over-the-air) configuration messages.

Here's a sample of the URLs where phone makers offer SMS configuration messages for e-mail, Internet browsing, and MMS. Check your phone-maker's Web site to determine if it offers SMS phone configuration:

✦ **Nokia:** www.nokia.com/support/

✦ **Sendo:** www.sendo.com/config/index.asp

✦ **Siemens:** www.siemens-mobile.com/mobiles (go to Customer Care and then Settings Configurator)

Configuring a phone via SMS is a straightforward process. Here is an example using the Nokia phone-configuration tool:

1. **Point your PC browser to Nokia.com (see Figure 1-6).**

2. **Select the Support tab from the tab options at the top of the page.**

3. **Select your region and country and click Go (see Figure 1-7).**

4. **Select a phone model from the Phone Model drop-down list and click Go (see Figure 1-8).**

Figure 1-6: Point your Web browser to Nokia.com.

Figure 1-7: Pick your geographical area.

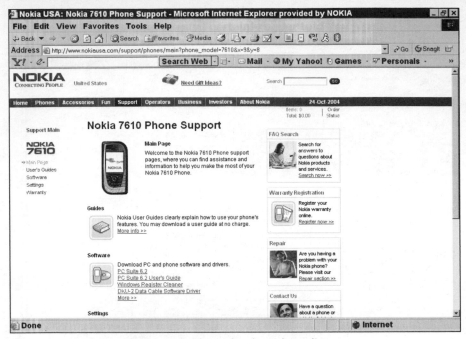

Figure 1-8: Select a phone model from the drop-down list.

5. **Click Settings.**

   The screen shown in Figure 1-9 appears.

6. **Under MMS Setup, click Start (see Figure 1-10).**

7. **Select your country from the drop-down list and click Go (see Figure 1-11).**

8. **Select your mobile network and click Go (see Figure 1-12).**

Figure 1-9: The settings screen.

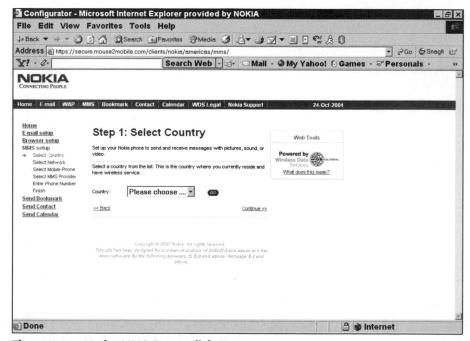

Figure 1-10: Under MMS Setup, click Start.

Figure 1-11: Select your country and click Go.

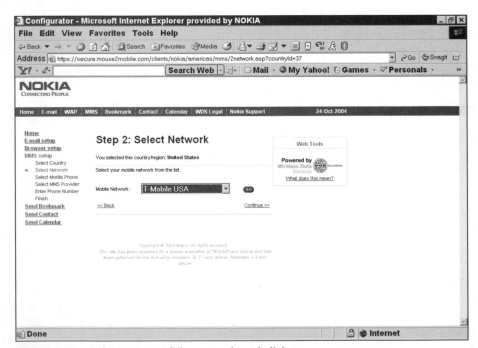

Figure 1-12: Select your mobile network and click Go.

9. Select your mobile phone and click Go.

10. Select your MMS provider from the drop-down list and click Go (see Figure 1-13).

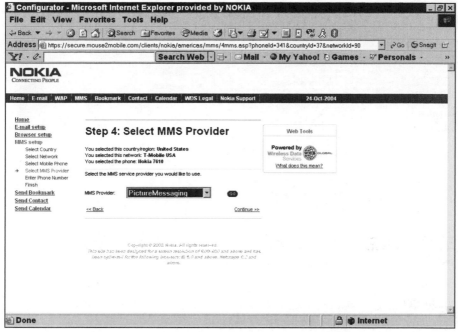

Figure 1-13: Select your MMS provider from the list and click Go.

11. Enter your phone number and click Send Settings (see Figure 1-14).

    Your phone receives a special configuration SMS message.

12. Go to the Inbox on your phone.

13. Select the new configuration message from your list of messages.

14. Press Options on your left menu key.

15. Choose Save Configuration Settings.

16. Press Yes at the confirmation screen.

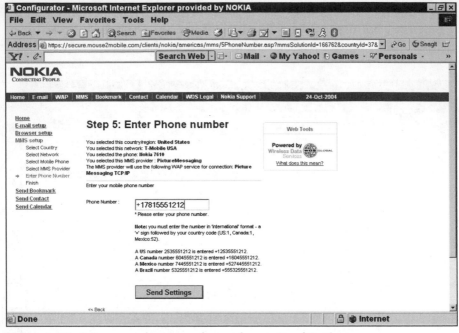

Figure 1-14: Enter your phone number and press Send Settings.

Congratulations! Your phone is now fully configured for sending multimedia (MMS) messages. And you did it yourself!

Technique 3: Moving Your Old Phone's Data to Your New Phone

A year or so ago, you wouldn't have wanted or needed to really worry about saving your old cellphone data and moving it to a newly purchased phone. Why? Two reasons:

✦ You really didn't have much data stored on your phone. Maybe a handful of speed-dial numbers in your phone book, but nothing you couldn't reenter in a few minutes by hand on your new phone.

✦ Your old phone really didn't offer any mechanism you could use to transfer information to and from the phone itself.

Both of these things have changed dramatically, and your smartphone has the capability to hold megabytes of information and the technology to share it with other devices (using several different communication technologies).

What you need to do depends on your situation. Because you're reading this book, I assume your new phone is a Series 60 phone (see the Introduction for information on how to tell if your phone is a Series 60 phone). So, this leaves two scenarios: Either your old phone is not a Series 60 smartphone or it's a Series 60 smartphone. Now, jump to the section that describes the scenario you need.

Moving data from a non–Series 60–based phone to your new Series 60–based phone

In this situation, your options are limited. Your old phone may or may not feature any of the communication ports — infrared, Bluetooth, or even a cable connection (USB) — that your Series 60–based phone may offer. If it doesn't, you won't be able to connect your old phone to a PC or to your new phone in order to transfer your information.

The good news is your old phone doesn't hold a lot of information that you'll want to transfer. Many older phone models can't store documents or files. So, the only items you may need to save would be contacts.

For older phones, your only real hope in saving your contacts is through SMS or text messaging. Some older phone models allow you to send contact information via an SMS message. Read your older phone's user guide for information on sending contacts (sometimes called business cards) through an SMS message. You can also scan through the menu options inside your older phone's contact application. Look for menu options like Send as Business Card or Send as SMS.

If you can send your contacts through SMS, you'll very likely be forced to do them one at a time. Here's the simple procedure:

1. **On your older phone, send an individual contact as a business card–style SMS to your new phone.**

2. **On your new Series 60 phone, on the main menu, select Messages.**

3. **Select Inbox.**

4. **Open the business card SMS you received from your old phone (see Figure 1-15).**

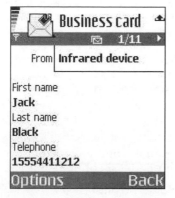

Figure 1-15: Your inbox will show the SMS from your old phone as a business card that holds special contact information rather than a regular SMS text message.

5. **Select Options from the left menu key.**

6. **Select Save Business Card (see Figure 1-16).**

   You'll see a confirmation message that says `Business card saved to contacts`.

Figure 1-16: Choose Save Business Card from the Options menu to automatically store the contact record from your old phone into your new phone's contact database.

7. **Select Options again.**

8. **Select Delete to remove the business-card SMS from your inbox.**

   This will make room for your other business cards coming from your old phone.

9. **At the Delete Message confirmation screen, select Yes.**

Repeat this procedure for every contact you need to transfer from your old phone to your new Series 60 phone.

Tip   Did you change service providers when you moved from your old phone to your new Series 60 phone? Believe it or not, you still may be able to use this technique, especially due to the opportunity for users to keep the same phone number when changing service providers through local number portability (LNP). Ask your service provider (or your new service provider if you've switched service providers) about sending SMS messages from your old phone to your new phone. Usually, there is a 12- to 24-hour period in which your old phone can make outgoing calls and send outgoing messages and your new phone will receive all incoming messages and calls directed to the same number.

Moving data from a smartphone to your new Series 60–based phone

In this scenario, you'll likely have several options. In the worst case, you could use the same SMS business-card technique described in the preceding section (moving data from a older phone model to your Series 60 phone), but you probably have faster, easier options available.

The most efficient method of transferring a large amount of contacts, files, and data is to synchronize your old smartphone to a PC or notebook computer (see Chapter 9 for details) and then synchronize your new Series 60 phone to the same PC. Using this approach, you can quickly select the specific files, contacts, images, and so on that you'd like to transfer to your new Series 60–based phone and—voila!—your Series 60 phone's PC suite software can copy it there.

Another option is to send individuals' contact records, files, or images directly from your old smartphone to your new Series 60 phone using a wireless connection (infrared or Bluetooth). This method is faster than the SMS technique from the preceding section because you don't need to go through your service provider's network. It's fast, simple communication between two phones. Here are the steps:

1. **On your old smartphone, navigate to the individual contact record, file, or image that you want to transfer to your new phone.**

   In this example, I'm sending an image from one phone to another, so on the main menu, I chose Gallery (choose Camera ➪ Images on the Siemens SX1).

2. **Select an individual image that you want to transfer.**

   Either press down on the joystick or choose Options ➪ Open from the left menu key.

3. **Select Options**

4. **Select Send (see Figure 1-17).**

Figure 1-17: Choose Send to transfer an individual image from phone to phone.

Tip Make sure you've activated infrared on your new phone. On your main menu, choose Connections ➪ Infrared.

5. **Select Via Infrared (see Figure 1-18).**

   The image appears automatically in your new phone's inbox.

Figure 1-18: Choose Via Infrared to send the image directly to another phone via a wireless connection.

6. **Select the image.**

7. **Select Options.**

8. **Select Save and pick the folder into which you want to store the new image.**

Technique 4: Registering with Your Service Provider's Web Site for Extra Goodies

Do you register your consumer electronic purchases — like a digital camera, camcorder, or DVD recorder — with the manufacturer? In the old days, product registration was all done exclusively via regular mail — you know, those little warranty postcards on which you write your product's serial number and the date of purchase and send it back to the company that made the product?

Now a lot of this type of product registration happens over the Web. And, in my experience, there are really only two types of people: those who diligently register their products and those who think about doing so but for whom important things — like a Red Sox game, lunch, playing with the kids, or learning new cellphone tricks — keep getting in the way.

I'm the latter type, in case you hadn't guessed. But even when all the important things come up, I still find time to register my cellphone with my service provider. Why? Because they frequently provide free ring tones and other downloads, as well as convenient ways to enable instant messaging and other cool features.

Every service-provider Web site allows its customers to register for free. Here is an example of the process using T-Mobile's Web site:

1. **Point your PC Web browser to T-Mobile's Web site** (www.t-mobile.com).

2. Select Register for My T-Mobile at the bottom of the page (see Figure 1-19).

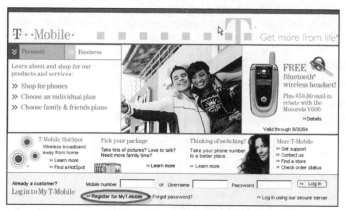

Figure 1-19: Select Register for My T-Mobile.

3. Enter your phone number in the Phone Number field and click submit (see Figure 1-20).

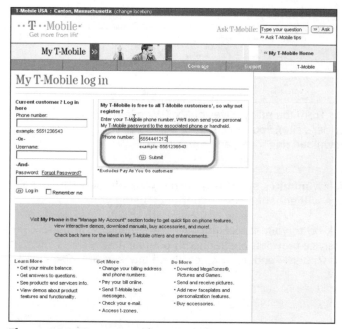

Figure 1-20: Enter your phone number.

4. Enter a user name and password for yourself.

5. **Click Submit.**

   You'll see the main screen of My T-Mobile (see Figure 1-21). From this page, you can check your minute usage, pay your phone bill, update your profile, request SMS news and entertainment updates, download ring tones and wallpaper, set up instant messaging, and much more.

Figure 1-21: From the many pages of My T-Mobile (or your service provider's site), you can pay your bill, check your minutes, download ring tones, and much more.

As you can see, a few minutes spent registering your phone with your service provider's Web site will enhance your experience using your phone.

Tip Log back on to your service provider's site regularly. In these competitive times, service providers are frequently offering new services and features to their subscribers—and many times, these offerings are inexpensive or free!

Turbocharging the Way You Make and Take Calls

If you've scanned through this book's table of contents and flipped through these pages, you know your smartphone can do some amazing things. Still, you're probably not surprised that the most commonly used features of smartphones are those that have to do with making and taking calls.

Fortunately, even in this area, your smartphone runs circles around your old cellphone. These Series 60–based phones (see the Introduction for an explanation of Series 60) offer several advanced features that help you make and receive calls more easily, more quickly, and in many cases, without using your hands!

If you want to get the most out of your smartphone, you need to know much more than the basics of calling described in your user manual. By learning these techniques, you'll be able to talk the talk when you want, with whom you want, like a cellphone grand master fourth-degree black belt.

This chapter shows you how to make voice-dialing work more effectively, make calls by pressing a single key, better identify who is calling you, manage your calls using some cool software downloads, and more.

Technique 5: Making Voice-Dialing Effective

Dialing your phone by voice is one of the more powerful features of your smartphone. This type of hands-free calling enables you to call while driving or otherwise engaged without endangering yourself or others.

Unfortunately, voice-dialing can also be a frustrating experience, particularly when the phone can't match your voice command to a particular contact record. Before I learned the tips in this section, I sometimes found myself shouting into my phone, "Call home! Call home!" only to look up and see a crowd of amused onlookers.

With these techniques, you can greatly improve the odds of getting a correct match. The ideas are simple, really. But these are the common-sense techniques that most people learn only after several mistakes. And, in the case of voice-dialing, most people give up before they find out what really works. I save you the time and energy of learning from your mistakes and let you learn from mine.

In order to improve your phone's voice-dialing capabilities, do the following:

✦ **Use your main microphone.** Microphones have different acoustics and levels of quality. If you record your voice tags with one microphone (like the phone's standard mic) but have trouble invoking the voice tag using another microphone (say a headset mic), the microphones themselves may be the culprit. So, record the voice tags using the same microphone you'll typically use to invoke the voice commands. This means if you'll primarily be using voice commands over a Bluetooth headset, then record the voice tags using the Bluetooth headset, not the standard phone microphone.

✦ **Be in the right place.** Record the voice tags in the same environment where you'll typically be when you use the voice commands. Acoustics of homes, offices, and cars vary tremendously. For example, if you plan to use voice commands frequently while driving, record the tags while sitting in your car.

✦ **Find a quiet moment.** When you get in the proper environment, try to record the tags during fairly quiet moments so that sporadic background noise isn't forever a part of your voice tags.

✦ **Think hard consonants!** Vowel sounds are hard to differentiate when matching audio clips. If possible, emphasize the consonants ("PeTri," "MiChael," "JaCKie"). If the name is all soft sounds ("Lisa"), try to add surrounding words or characters to make the voice tag stand out ("Call Lisa K.").

Employ these techniques and you'll find voice-dialing can be a powerful hands-free productivity enhancer.

 Tip Not only is dialing by voice a good idea — in your state, it may be the law! Already in New York and coming in Massachusetts, driving while using a cellphone is illegal — unless you're using a car kit or headset. And the best way to place calls using these cellphone accessories is by voice.

The way to make voice-dialing work on a Series 60–based smartphone is to create a voice tag and assign it to an individual contact's number. Here are the steps to follow:

1. **Open up a particular contact record in Contacts (see Figure 2-1).**

Figure 2-1: Open up an individual contact.

2. **Choose Options ⇨ Add Voice Tag (see Figure 2-2).**

   The phone displays `Press Start, then speak after the tone.`

Figure 2-2: Choose Options ⇨ Add Voice Tag.

3. **Choose Start and speak after the tone.**

   Note   The Siemens SX1 will ask you to repeat the voice tag a second time for each contact.

4. **Press Quit to view the contact card after you successfully make a recording.**

Sendo's voice-recognition technology

The Sendo X does not use voice-tag technology. Instead, it uses speaker-independent voice-recognition technology to understand and execute what you're asking it to do. This means you don't need to prerecord voice-tag information for your contact records.

The Sendo X has special settings to tune this voice recognition to better suit your voice and your use of the phone. Here's how to access the Sendo X's voice settings:

1. **On the main menu of the Sendo X, select the Tools icon.**

2. **Scroll to Settings and select it.**

3. **Scroll to Voice Settings and select it.**

On this screen, you'll see Choice Lists and Recognition settings.

4. **Adjust these settings to best suit your manner of voice-dialing.**

Additionally, the Sendo X has a voice key on the side of the phone. Just press it and say the command "name dial." Then, follow the prompts to call a particular contact.

To activate your new voice tag, in standby mode, hold down the right menu button and when you hear the beep, say the voice tag. Your phone will automatically dial the contact associated with the voice tag.

Tip Nokia phones store up to 30 voice tags. Only one voice tag can be associated with a particular contact record, even if you have more than one number listed for that contact. To view all voice tags on your Nokia phone, choose Options ➪ Contacts Info ➪ Voice Tags in your Contacts directory.

Technique 6: Making Calls by Pressing Only One or Two Keys

Speed-dialing works on both regular phones and cellphones. It's an easy, fast way to dial your most frequently called numbers. Just as with your home phone or your office phone, on your Series 60–based smartphone, you can assign your most frequently called numbers to the number keys 2 through 9. On most cellphones, the 1 key stays reserved for your voicemail number or an emergency number like 911 across the U.S. and 112 in Europe.

Note On the Sendo X, keys 2 and 3 are also reserved. The 2 key activates the WAP browser and the 3 key activates the camera.

Note
On all Series 60/Symbian phones, you can assign a new 1-Touch key quickly and easily while in Standby mode. Simply press an unassigned number key (like the 4 key, for example) and then press the call button.

If you press an unassigned speed-dial key and the send button, you'll be asked if you want to assign that key to a contact (see Figure 2-3).

Figure 2-3: Pressing an unassigned speed-dial and the send button prompts the phone to ask if you want to assign it.

The screen will ask if you want to assign the number now. If you select Yes, the Contacts Manager will open up. You can scroll to the name of the contact you want to assign to that number key, then choose Select. Shazam! You've set up a new speed-dial key.

Assigning a speed-dial key

To assign a speed-dial key for all Series 60/Symbian phones, follow these steps:

1. **On the main menu, scroll to Tools (Settings on the Siemens SX1) and select it.**

2. **Scroll to 1-Touch (or Speed Dial or Shortcuts depending on your phone's icon names) and select it (see Figure 2-4).**

Figure 2-4: Select 1-Touch (or Speed Dial or Shortcuts).

3. **Scroll to an unassigned key (all assigned keys will have an icon or contact name associated with them) and select it (see Figure 2-5).**

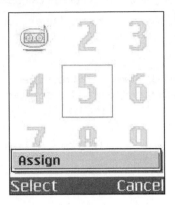

Figure 2-5: Highlight an unassigned 1-Touch key and choose Select ⇨ Assign.

4. **Select the Assign option.**

5. **Browse to the contact you want to associate with that speed-dial key and select it.**

   If you have more than one phone number for that contact record, you'll be asked to select one of them.

Using the standard two-key method of activating speed-dial

To use the standard two-key method of activating a speed dial for all Series 60/Symbian phones, follow these steps:

1. **In Standby mode, press the speed-dial (number key) you want to call.**

2. **Press the Send button.**

Turning on 1-Touch dialing

Here is the cool technique and one you would have to read the fine print of your user manual to find. You can turn on 1-Touch dialing, which will allow you to press only one key, as the name implies, to call one of your speed-dial contacts.

To turn on 1-Touch dialing for all Series 60/Symbian phones, follow these steps:

1. **On the main menu, select Tools.**

 Note Settings is found on the main menu, not the Tools folder, on some Series 60 phones.

2. Select Settings.

3. Select Call Settings.

4. Scroll down and select 1-Touch Dialing.

5. Select On (see Figure 2-6).

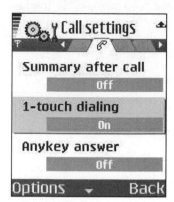

Figure 2-6: Turn on 1-Touch dialing.

6. Select Back.

7. Select Exit.

Now, in standby mode, simply press and hold an assigned speed-dial key for a few seconds and — *voilà!* — you're making a call.

Technique 7: Creating Groups for Faster Identification and Communication

Many phone users know you can assign specific ring tones for individuals in your contacts list. This way, using caller ID, you can identify who is calling just by listening to the ring.

Here's the cool twist to that feature. By categorizing your contacts into groups, you can not only quickly separate work-related callers from personal or family callers, you can also empower yourself to send text messages to many people simultaneously.

Creating a new group of contacts

To create a new group of contacts, follow these steps:

1. **On the main menu, select Contacts.**

2. **Move to the right with the joystick until you see your list of groups.**

   If you currently have no groups, you'll see "(no groups)" in the center of the screen (see Figure 2-7).

Figure 2-7: View your list of contact groups.

3. **Choose Options.**

4. **Select New Group.**

5. **Enter and name your group (see Figure 2-8).**

Figure 2-8: Enter a new group name.

6. **Select OK.**

Assigning a ring tone to a group

Now that you've created a group, let's assign a different ring tone to the group so that you can easily tell when anyone in the group calls you:

1. **In the group section of Contacts, use your joystick to highlight a group.**

2. **Select Options.**

3. **Select Ringing Tone (see Figure 2-9).**

Figure 2-9: Select Ringing Tone to assign a particular ring tone to a whole group of people.

4. **Scroll through your list of ring tones.**

Note   What do those ring-tone filename extensions like AMR and MID mean? AMR is a mobile-phone-specific audio format, so ring tones with an AMR extension are most likely to be real recorded sound—like the popping of a champagne cork or the blast from a steam locomotive—that has been converted into a ring tone. MID is the extension for a MIDI sound file. Ring tones with an MID extension will most likely be *polyphonic* tunes (with many notes at once). These might be mimicking popular songs, classic songs, or just newly created ring-tone tunes. If a ring tone has no file extension, it will most likely be *monophonic* (only one note at a time), which is usually a less impressive version of a song than a polyphonic ring tone of the same tune.

By pausing on a particular ring tone as you scroll through the ring-tone list, your phone will begin playing the ring tone, giving you a sample of what it would sound like if you selected it as your phone's tone.

5. **Highlight the ring tone you want to assign to your new group and press Select.**

Adding members to your group

Now that you've created a new group and assigned a ring tone, you'll need to add members to it:

1. **Go to the Groups section of your Contacts list.**

2. **Highlight a particular group.**

3. **Select Options.**

4. **Select Open.**

5. **Select Options.**

6. **Select Add Members.**

   If you only want to add one member, simply highlight the individual contact and select OK. To add more than one member at a time, scroll to each contact you want to add and center-press on the joystick. This action will mark the check box next to that contact's name (see Figure 2-10).

Figure 2-10: Check the box next to a member to add the member to your new group

7. **When you're done adding members, select OK.**

 Tip You can remove a check next to a contact by center-pressing the joystick while the contact is highlighted. Center-pressing on the joystick toggles between checked and unchecked in your Contacts list.

 Tip Is there a way to tell if you've assigned a contact to more than one group? You bet. In Contacts, highlight an individual name, choose Options, and select Belongs to Groups. You'll then see a list of all the groups to which that individual has been assigned (see Figure 2-11).

If an individual contact is assigned to more than one group, what ring tone is played when that individual calls you? Whatever ring tone was last assigned to any group associated with that individual. So, say your sister belongs both to Family and Personal groups. If you last created the Personal group and added your sister, then the ring tone assigned to the Personal group will be the one played when your sister calls. If, however, you then go and change the ring tone assigned to the Family group, then that new ring tone will be the one played when your sister calls because it was the latest ring tone updated that's associated with your sister's contact record.

Confusing? A simple way to check is to open the individual contact record. There you'll find a ring tone listed. That's the ring tone that will play when the person calls. It gets automatically updated when you assign or change group ring tones.

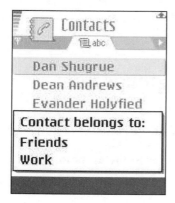

Figure 2-11: Choosing Options ➪ Belongs to Groups shows you the list of all the groups to which an individual contact belongs.

Sending a message to a group from the Group List screen

The ability to send a single message to a group means you can easily tell all your work colleagues the latest news from a meeting or tell your whole circle of friends you'll be late for the party. Follow these steps:

1. **In the main menu, select Contacts.**

2. **Using the joystick, scroll right until you come to the list of groups.**

3. **Highlight the group to which you want to send a message.**

4. **Select Options.**

 Tip   When a group is highlighted, pressing down on your joystick opens the group.

5. **Select Open.**

6. **Select Options.**

7. **Choose Create Message ➪ Text Message (or Multimedia Message — whichever type of message you want), as shown in Figure 2-12.**

Your message editing window will appear with all the members of the group automatically entered in the To field.

Figure 2-12: You can create a group message right from the groups list.

Addressing a message to multiple groups from the Edit Message screen

To address a message to more than one group, follow these steps:

1. **In the message editing window, move the cursor to the To field.**
2. **Select Options.**
3. **Select Add Recipient.**
4. **In the Contacts screen, scroll right using the joystick to get to the list of groups.**
5. **Center-press on the joystick to check the box or boxes of the groups to which you want to send the message (see Figure 2-13).**

Figure 2-13: Check the boxes for multiple groups to send your single message to a wide audience.

6. **Select OK when you're done.**
7. **Finish and send the message as you normally would.**

Now with a single message, you can tell all your friends to meet in a particular spot, inform all your kids that you're ready to leave the shopping mall, or alert your colleagues that you've changed the meeting room for the upcoming meeting.

Technique 8: Adding an Image to Your Contact to Better See Who Is Calling You

You've heard that a picture is worth a thousand words. Now, medical research seems to be saying that pictures may move faster through your brain than words as well—at least in terms of memory triggers and recognition.

Save yourself a few nanoseconds every time someone calls by inserting pictures into individual contact records. That way, you'll recognize the person by sight instead of having to read his name.

To add an image to a contact record, follow these steps:

1. **In the main menu, choose Contacts.**
2. **Scroll or search for an individual contact (see Figure 2-14).**

Figure 2-14: An individual contact record with no thumbnail image.

3. **Select Options.**

4. **Select Edit.**

5. **Select Options.**

6. **Select Add Thumbnail (see Figure 2-15).**

Figure 2-15: Select Add Thumbnail.

A Select Memory pop-up window appears.

7. **Choose Phone Memory or Memory Card, whichever one holds the image you want to insert for this contact (see Figure 2-16).**

Tip For the Siemens SX1, there's another way to add an image to a contact record. It's called Image View. To add an image to a contact on the Siemens SX1, in the main menu select Contacts. Highlight an individual contact, select Options, and select Open. Move the joystick to the right to open the Image view and then select Options and select Add Image. Navigate to the image you want to add, and select OK.

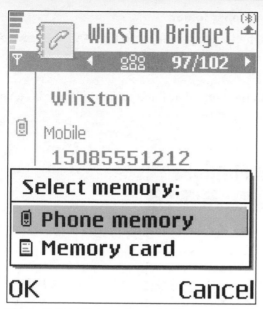

Figure 2-16: Choose the memory area where your image resides — phone or memory card.

8. **Select OK.**

The Select Image screen appears, showing all the folders within the memory area you selected in the last step (see Figure 2-17).

Figure 2-17: Select the image you want to insert into the contact record.

9. Navigate down into the proper folder.

10. Scroll to and highlight the image you want to insert in the contact record.

11. Choose Select.

   The image will appear in the upper-left-hand corner of the contact's record (see Figure 2-18).

Figure 2-18: The image appears in the upper-left-hand corner of the contact record.

Now, when this contact calls you, you'll see her picture as well as the contact name.

 Note   Contact images also show up in your 1-Touch Dialing screen, replacing the name of the contact that you've assigned to a particular number key.

Technique 9: Looking Up Your Voicemail Number

Quick, what's the phone number of your camera phone's voicemail box? You may not know what it is. Many people don't. Most cellphones come preprogrammed from the carrier with the voicemail number. If yours is, you access your voicemail through a speed-dial setting or by entering a simple shortcut code, like 1-2-3.

So, what's the problem? Well, if you ever want—or need—to access your smartphone's voicemail from some other phone, doing so would pose a problem. What happens if you temporarily misplace your smartphone, but you still need to get your voicemail messages?

No worries. This little trick will let you see your phone's voicemail number. Write it down someplace—other than your smartphone—so that you can get it if you need it.

To look up your voicemail number, follow these steps:

1. **In the main menu, select Tools (Settings on the Siemens SX1).**

2. **Select Options.**

3. **Select Open.**

4. **Select Voicemail.**

5. **Select Options.**

6. **Select Open.**

   Here, you'll see the number for your voicemail box (see Figure 2-19). Note it down somewhere safe.

Figure 2-19: Ta-da! Observe your preconfigured, somewhat hidden, voicemail phone number.

 If you call your voicemail from a phone other than your cellphone, you may need to enter your smartphone's number, press the * key, and enter your voicemail password, if you've set one, before you can access your messages.

Technique 10: Fooling Your Callers with Fake Background Noises Using the CallCheater Download

On the upside, cellphones enable you to be in touch anywhere at any time. On the downside, cellphones enable you to be in touch anywhere at any time. That's right, having a phone with you at all times is both a blessing and a curse.

Even the most noble among us have moments when we need privacy, perhaps when we're someplace where at least one person on earth doesn't think we should be. For example, my boss wouldn't take it so well if she knew I was at a Red Sox game during working hours (not that I've ever done that, of course). My wife wouldn't be very understanding if she were to call me and hear the noise of a bar in the background when I was supposed to be driving home to watch the kids.

The good news: A shareware application called, appropriately, CallCheater can help with these types of situations. CallCheater simply plays background noises that you select as appropriate for your caller to hear.

I'm not suggesting that you use CallCheater to lie to your friends, family, and business colleagues. But I'm making you aware of yet another example of powerful application shareware available over the Web. Don't try this with some lesser cellular platform offering—the Series 60 platform's multitasking capability is what makes this all work. Your phone can run an application while you talk on the phone. Most phones can't do this (including many based on "advanced" operating systems) because they don't allow for multiple applications to interact and stay running simultaneously.

CallCheater is available to download from SymbianWare (www.symbianware. com). Read Technique 72 for instructions on installing Symbian software. **Remember:** As with any shareware application, try it before you buy it. CallCheater offers a 15-day trial version that will disable itself after your 15 days expire.

After CallCheater is installed, here is a quick description of how to set it up:

1. **On the main menu, select CallCheater.**

   Note: All new Symbian application installations are installed to your phone's main menu. If you want, you can move the application to a different folder using Technique 37.

2. **Select Sounds on CallCheater's menu (see Figure 2-20).**

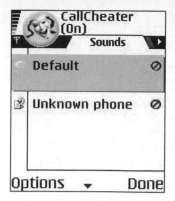

Figure 2-20: Select Sounds on CallCheater's menu to start your initial setup.

3. **Select Default (see Figure 2-21).**

4. **Choose Set Sound on the pop-up menu (see Figure 2-22).**

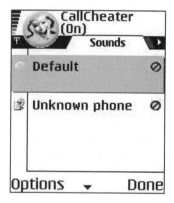

Figure 2-21: Select Default to choose a sound to play when people call you.

Figure 2-22: Choose Set Sound to see a list of sounds to assign to this sound rule.

5. **Choose a sound from the Select Sound list that comes with CallCheater.**

   As you scroll through the list using the joystick, you'll hear the sounds played. You can choose from traffic noises, garden sounds, weather, and office-equipment racket.

6. **Back on the Sounds tab, you'll see your selected sound name appear under Default (see Figure 2-23).**

7. **Select Default again.**

8. **Choose Enable/Disable from the pop-up list (see Figure 2-24).**

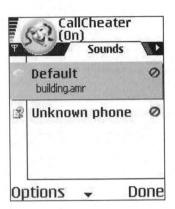

Figure 2-23: You'll see the sound name you selected under Default.

Figure 2-24: Choose Enable/Disable in order to enable your new default sound rule.

9. **Choose Enable (Pen+Right).**

   A checkmark will appear next to your new enabled Default sound rule. But, you're not done yet.

10. **Select Done on the right menu key.**

11. **Select Activate on CallCheater's menu (see Figure 2-25).**

Now, when you receive a call, both you and your caller will hear the background noise you selected.

CallCheater has many other options that you can explore on your own and by reading the CallCheater documentation that you can download along with the trial software from SymbianWare.com. You can, for example, create new sound rules that apply to specific contacts from your contact directory—like always play office noise when your boss calls you. You can set CallCheater to work automatically when you receive a call or you can set it in manual mode, where you must activate CallCheater using your joystick while you're taking a call. You can even add your own background noises from sound files you've created.

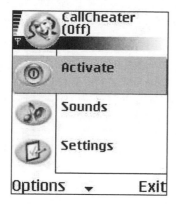

Figure 2-25: Select Activate to start CallCheater using your new rule.

Technique 11: Automatically Rejecting Unwanted Calls Using the BlackList Download

In the early days of wireless, cellphones were off-limits to telemarketers—mainly because people were and still are charged for every minute of time on the phone, even when receiving calls. But as Bob Dylan would say, "The times, they are a changin.'" Now even your cellphone can be a target for annoying telemarketers. Fortunately, once again, the shareware developers have come to your rescue.

Using the BlackList download, you can specify exactly whom you want to receive calls from. All other callers will be automatically rejected without your having to touch a single key.

BlackList is available from SymbianWare.com. (Read Technique 72 for instructions on installing Symbian software.) As always, try it before you buy it. BlackList, and all the other shareware found on SymbianWare.com, offers a free 15-day trial version.

After you've installed BlackList, here's an example of how to set it up. In this example, I show you how to set BlackList to always reject one particular annoying caller. You can also set BlackList to reject every call *except* the list you specify. So, it can be used in either a positive (everyone gets through but this list) or negative (the only people who get through are on this list) way. Read the documentation that you can download along with the trial version for instructions on configuring BlackList.

To block a particular caller whose number you know:

1. **On the main menu, select BlackList (see Figure 2-26).**

Figure 2-26: Select BlackList.

Inside the application are three tabs: Status, Rules, and Rejected.

2. **Press the joystick to the right to go to the Rules tab.**

3. **Select Black List (see Figure 2-27).**

Figure 2-27: Select the Black List rules.

You'll see a red dot appear next to the Black List.

4. **Select Options on the left menu key**

5. **Choose Edit List (see Figure 2-28).**

6. On the BlackList correspondents list, select Options on the left menu key.

7. Choose Add Person (see Figure 2-29).

Figure 2-28: Choose Edit List to add a caller to your automatic rejection list.

Figure 2-29: Choose Add Person to add a contact to your BlackList.

Your contact directory will appear.

8. Scroll to the name of the person whose calls you want to block.

9. Select the name.

The contact will appear in your correspondents list (see Figure 2-30).

10. Select Done on the right menu key.

11. Press the joystick to the left to move to the Status tab.

12. Select Turn On in the right menu key (see Figure 2-31).

Now when your black-listed caller rings your phone, BlackList automatically rejects him without your doing a thing. On the BlackList Status tab, you'll see a running tally of rejected callers (see Figure 2-32). The Rejected tab shows you exactly whose calls were blocked.

Figure 2-30: The contact appears in the correspondents list.

Figure 2-31: Select Turn On to activate your BlackList.

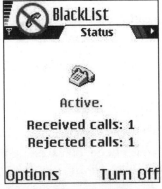

Figure 2-32: BlackList keeps a tally of rejected calls.

Technique 12: Sending Automatic Text Message Responses to Callers Using the SMSMachine Download

Wouldn't it be nice to have a personal assistant who would notify your callers that you're in a meeting, traveling, or at lunch but will return the call as soon as you get back?

These days many e-mail systems have an automated reply feature that lets you automatically send a reply e-mail to people who send you a message, telling them that you're currently unavailable. In the automatic reply message, you can specify that you're on vacation, traveling on business, in all-day meetings, or whatever.

With phones, you could try to use your voicemail greetings to perform this same function, but it just isn't as convenient and you can't create specific greetings for specific callers (at least on most voicemail systems).

Tip You can make people aware of your availability with something called Presence. Through the Presence feature, friends, family, colleagues, and so on can subscribe to your "presence," allowing them to see if you're available for a phone call simply by looking at your entry in their phone's contact directory. Whenever you go into a meeting, leave work, or travel, you quickly update your own presence setting and people who subscribe to your presence will automatically be updated. Series 60 phones support Presence, but your service provider also needs to have a Presence server that you and those you interact with use. Presence servers are rolling out around the world through service providers. If you're interested in Presence, ask your service provider if it supports it.

Never fear though, because another shareware program neatly adds this capability to your Series 60 phone. It's called SMSMachine, and it operates like an out-of-the-office wizard or a personal assistant for your phone.

You can download SMSMachine from SymbianWare.com (`www.symbianware.com`). Read Technique 72 for instructions on installing Symbian software.

Like most of the shareware applications mentioned in this book, SMSMachine can be configured in various ways. Here's an example of how to set up an "I'm in a meeting" automated response:

1. **On the main menu, select SMSMachine (see Figure 2-33).**

   Inside SMSMachine, you'll see three tabs: Main, Responses, and Log.

2. **Press the joystick to the right to move to the Responses tab.**

   Two responses come preloaded — default and black list.

Figure 2-33: Launch
SMSMachine.

3. **Select Options on the left menu key.**

4. **Choose Add New Response (see Figure 2-34) to create a new type
   of automated response message for your callers.**

Figure 2-34: Choose
Add New Response.

5. **Choose Response Name (see Figure 2-35).**

6. **Enter** Meeting **(see Figure 2-36).**

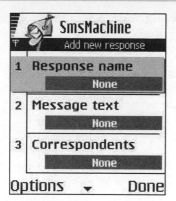

Figure 2-35: Choose Response Name to modify the name of the new automated response.

Figure 2-36: Enter a new response name.

7. **Scroll down to Message Text.**

8. **Select Message Text (see Figure 2-37).**

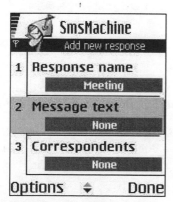

Figure 2-37: Select Message Text.

9. **Enter an appropriate message for alerting your callers that you're in a meeting (see Figure 2-38).**

Figure 2-38: Enter a message that will be automatically sent to your callers alerting them to your current status.

10. **Scroll down to Autoreply On.**

11. **Select Autoreply On.**

12. **Choose Both so that SMSMachine will send this message both to those who call and those who SMS you (see Figure 2-39).**

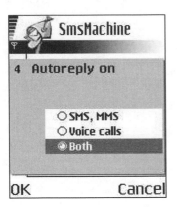

Figure 2-39: Select Both so that everyone who tries to contact you will be notified of your status.

13. **Press Done on the right menu key.**

14. **Meeting will now appear in your list of Responses (see Figure 2-40).**

15. **Select Meeting.**

   This will mark your current status with a checkmark (see Figure 2-41).

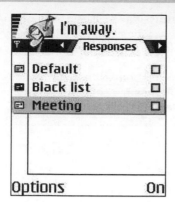

Figure 2-40: Meeting is now included in your list of responses.

Figure 2-41: Select Meeting as your current status. Your Meeting message will now be sent automatically to those who call or SMS you.

16. **Press the joystick to the left to return to the Main screen of SMSMachine.**

17. **Press Turn On on the right menu key (see Figure 2-42).**

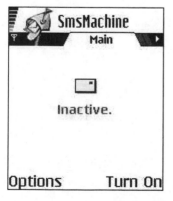

Figure 2-42: Turn on SMSMachine by pressing the right menu key.

The Main screen will show the word "Active" as well as a tally of received events and sent responses. Now, when you're called or SMSed, people will be automatically told you're in a meeting.

The best use of SMSMachine is to build a set of responses that suit your lifestyle with appropriate messages for each. A possible set would include: At Lunch, In a Meeting, Traveling, On Vacation, Out Sick, and so on. Isn't it nice to have a personal assistant on your smartphone?

Capturing, Saving, Editing, and Sending Images

CHAPTER

3

This book might have been titled *101 Cool Camera Phone Techniques* because every Series 60 phone — except the two Nokia game decks, the N-Gage and N-Gage QD — feature integrated digital cameras. But the savvy editors at Wiley knew that the term *camera phone* didn't really embody all that these smartphones can do.

The integrated digital cameras are the most obvious and most celebrated difference between older cellphones and the newer, advanced phones described in this book. In the course of a year, camera phones have impacted our world in a major way. Camera phones have caught criminals in the act; been banned from many gyms, locker rooms, and dressing rooms; and negatively impacted the sales of traditional digital cameras. The camera phone may be fundamentally changing the way that people capture and share images — and you're among the very first generation of people to catch this potentially tsunami-size wave!

This chapter takes you deeper into the cool ways to capture and share images with your Series 60–based camera phone. Your phone's manual only glosses over how to snap pictures and use images to customize your phone. I know because I've read all of them.

In this chapter, I show you how to capture an image with only two key clicks, rotate and zoom images, use a built-in timer for group shots, upload pictures to online photo services, and more.

One special note: As I've said, two notable Nokia phones — The N-Gage and the N-Gage QD — do not feature integrated cameras and, therefore, cannot accurately be called camera phones. However, both of these devices are Series 60–based phones and can perform every tip and technique described in these pages — except for capturing an image via an integrated camera. Please keep in mind that you can still get images onto the N-Gage and N-Gage QD phones in a number of ways, including MMS messages or via a Bluetooth or infrared connection to a PC. Keep in mind that after the images are on the N-Gage devices, you can perform the rest of the image-related techniques in the ways described in this chapter, just like every other Series 60 phone.

Technique 13: Setting Your Camera Phone for the "Fast Draw"

One of the main differences between using a traditional camera — either digital or film-based — and a camera phone is spontaneity. People like camera phones because they're always present and ready for action. You can snap the picture of your child's first steps or that perfect sunset because none of the purposeful planning and setup of using a traditional camera is required.

Camera phones are even more convenient and more likely to be on hand than the small point-and-shoot digital traditional cameras that have been very popular for the last several years. So with spontaneity in mind, let's make sure your camera phone is set for immediate picture taking or, as I sometimes call it, the "fast draw."

Because your camera phone will most often be in standby mode (see "Terms you need to know" in the Introduction), you want to set the left or right menu key to launch the camera function. This way, with one click, you'll be ready to capture that moment and keep it forever.

Activating your camera by voice

Say you want fast access to the camera but don't want to assign one of your standby-mode menu keys to the camera. There are other ways you can quickly start the camera.

Read your phone's user manual to see if it supports Voice Command (which lets you start applications, not just dial your contacts by voice). If your phone offers the Voice Command feature, you can assign a voice command to put your phone in camera mode.

To assign Voice Command to launch camera mode, follow these steps:

1. **From the main menu, open Tools.**

2. **Select Voice Command (see the figure).**

Select Voice Command.

3. **Highlight Camera on the list of applications (see the figure).**

   If you don't see Camera listed, choose Options. Then select New Application and choose Camera from the full list of applications.

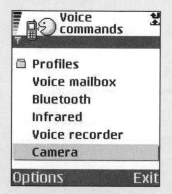
Select Camera.

4. **Select Add Voice Command.**

   You'll see `Press "Start", then speak after the tone.`

5. **Press Start.**

6. **Say your voice command.**

   Choose something you'll remember, like "Start Camera."

To activate your new camera voice command, in standby mode, hold down the right menu button and, when you hear the beep, say the command. Then you're ready to snap that photo.

Follow these steps to assign the camera function to one of the standby-mode menu keys (for all Series 60 phones except the Sendo X—for the Sendo X see the "Assigning a menu key on the Sendo X" sidebar):

1. **On the main menu, select Tools.**

Note On some phone models, like the Nokia 7610, the Settings folder is on the main menu and not in the Tools folder.

2. **Choose Settings.**

3. **Select Phone.**

4. **Choose Standby Mode (see Figure 3-1).**

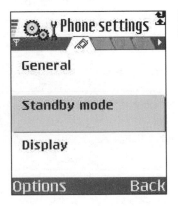

Figure 3-1: Choose Standby Mode.

5. **Select the Left Selection Key or the Right Selection Key (see Figure 3-2), whichever one you want to assign to the camera.**

Based on my own experience, for the very fastest draw, right-handers should choose Left Selection Key and left-handers should choose Right Selection Key. This enables easy, comfortable, one-hand activation of the camera.

Figure 3-2: If you're right-handed choose Left Selection Key for the fast-draw camera. Left-handers should choose Right Selection Key.

6. **From the list of functions, choose Camera (see Figure 3-3).**

Figure 3-3: Assign Camera to your menu key.

Tip

You may notice some items missing as you scroll through the left or right selection key lists. Why? Anything that is already assigned to the other menu key won't be available. For example, if you have Contacts assigned to your left menu key, it won't appear on the right selection key list.

7. **Press OK.**

8. **Choose Back to go to Phone Settings.**

9. **Choose Back to go to Settings.**

10. **Choose Exit to go back to the main menu.**

Now, in standby mode, you'll see Camera ready and waiting for instant picture taking with the press of just one menu key (see Figure 3-4).

Figure 3-4: The fast-draw Camera function is ready for instant picture taking.

Assigning a menu key on the Sendo X

The Sendo X has a slightly different method for assigning functions to menu keys. The Sendo X has a special Now! screen that overlays the main menu. The Now! screen features different panes that you access by pressing the joystick left and right. The Sendo X allows you to create new panes and add applications to them.

To assign the camera function to one of the standby-mode menu keys, follow these steps:

1. On the Now! Screen, choose Options.

2. Choose Pane Settings.

3. Choose Right Selection Key.

   Note: The main screen on Sendo does not have Left Selection Key.

4. Select Camera.

5. Choose OK.

At this point, the Now! screen shows Camera ready and waiting for spontaneous picture taking.

The Siemens SX1 has a special camera function button on the right side of the phone, just below the SX1's voice command button. Press it once to activate the SX1's camera. Press it when the camera is active to capture an image.

If for some reason you don't want to assign the camera function to your menu keys, you can reorder your main menu so that the camera is always the first item in the menu. This move will allow you to launch the camera with two clicks—one click to go from standby to the main menu and one click to launch the camera.

To move the camera item to the first item in your main menu, follow these steps:

1. **From standby mode, press the main menu key to open the main menu.**

2. **Use the cursor to highlight the Camera.**

3. **Choose Options.**

4. **Select Move (see Figure 3-5).**

5. **Use the joystick to highlight a different menu position for the camera function.**

   For fastest access, choose the upper-left-hand corner.

6. **Choose OK to place the item on the menu in the new position.**

The menu will now be rearranged.

Figure 3-5: Select Move to rearrange items on your main menu.

Tip

One more thing about fast picture taking that you may not find in your user manual: When your camera is launched, you can capture an image using the Options menu (Options ➪ Capture = two clicks) or using one click of the joystick. Simply press down once on the joystick while in camera mode to snap a picture. Thus, in standby mode, with this "fast draw" setup, you can be only two clicks away from a captured image: one click to activate the camera (using the Camera menu customization described in this technique) and one press of the joystick to capture the image.

Technique 14: Zooming and Rotating Shortcuts While Shooting and with Saved Images

The earliest camera phones offered only minimal camera features — basically, capturing and saving an image. Phone engineers, however, have quickly followed up with camera phones that include other popular digital-camera features, like zooming and rotating images.

All Series 60 phones let you zoom and rotate within an image file. Newer Series 60 phones also allow you to zoom during image capture. Good luck finding any of these features on a non–Series 60 camera phone.

Using the Sendo X built-in camera flash

The Sendo X has another cool camera feature—a built-in flash! To activate the Sendo X flash, follow these steps before you snap the picture:

1. **Activate the camera (by choosing Camera from the main menu or, if you've followed Technique 13, by simply pressing the left or right menu button to activate the camera with one click).**

2. **Choose Options.**

3. **Select Activate Flash.**

Now, shoot your picture as you normally would. The flash will light a few seconds before the image is captured and stay shining until the capture is completed.

Sendo's flash will stay active—you'll see a lightning bolt at the bottom of the camera screen when the flash is active.

Here is how to zoom in and out and rotate a saved image:

1. **On the main menu, select Gallery.**

 Note Gallery is an option only in newer Series 60 models. Models like the Nokia 3650 have an Images folder on the main menu (on the Sendo X it's called Photos; on the Siemens SX1 it's called Images but it's in the Camera folder). Gallery is a newer application that consolidates all saved media files (images, video, audio, and so on). The old Images folder housed images only.

2. **Select a saved image, either from phone memory or your memory card (using the joystick to move left and right between phone memory and your memory card).**

3. **After the image is open, press the 5 key to zoom in.**

   Each click increments the zoom magnification by 25 percent. The maximum zoom magnification is 100 percent. Your current level of magnification is shown at the top of the display. You can also use the menus to zoom in by choosing Options ⇨ Zoom In, but this shortcut is much quicker and requires fewer key clicks.

   Note If you're viewing an animated GIF, zooming will temporarily stop the motion. The animation will resume after you return to normal view.

4. **To zoom out, press the 0 key.**

   It works just like the 5 key. Each click decrements the zoom magnification by 25 percent. You can also use the menus to zoom out by choosing Options ⇨ Zoom Out, but what fun is that?

Tip Press and hold the 0 key, and the image will return to its normal view.

Note Zooming in and out on a saved image is not saved.

Tip At any point while you're rotating or zooming, you can view the image in full-screen mode by choosing Options ⇨ Full Screen. Full-screen view is probably the best way to present your pictures to others at cocktail parties, at sporting events, during school detention, and at other social gatherings.

5. **To rotate an image counterclockwise, press the 1 key.**

   Each click rotates the image 90 percent. The image will keep rotating as long as you keep clicking, so if you want to spin the image — right round, baby, right round, like a record, baby, right round round round — go ahead and knock yourself out.

6. **To rotate the image clockwise, press the 3 key.**

   It also rotates 90 percent with each click.

Note As with zooming, you can use the menus to rotate by choosing Options ⇨ Rotate. (But that slow method is best left for the unfortunate lot that didn't buy this cool book!) Like zooming, rotating an image does not change how it is saved.

Technique 15: Setting Different Resolutions

Like many traditional digital cameras, your Series 60–based camera phone allows you to set different resolutions for the images you capture. Lower resolution means fewer digital pixels make up your image. Thus, low-resolution images look blurry or fuzzy when made large, yet they take up very small amounts of memory and they appear fine on small screens like typical cellphone screens.

The earliest camera phones offered maximum resolutions much lower than state-of-the-art digital cameras. The first camera phones, from various manufacturers, maxed out well below 1 megapixel, while consumer-level digital cameras currently feature up to 5 or 6 megapixels. Quite a world of difference!

This first wave of camera phones made no attempt to replace digital cameras with their low-resolution images. Instead, they were developed to allow users to simply "capture the moment." But times are changing quickly — so quickly, in fact, that you'll need to check the reference guide of your particular phone in order to see exactly what your camera phone's resolution capability is.

Table 3-1 is a sample chart from Nokia 6600 (lifted from the user manual of the phone) showing how many images of its three different types — standard, night, portrait — at its three different resolution settings — basic, normal, and high — will fit into 1MB of phone memory. The best possible pictures result from setting the phone to shoot Standard type images at the High resolution setting. Also, Portrait takes up the least amount of memory, although the quality of the images is not very high.

Table 3-1 Image Quality			
Picture Type	*Basic*	*Normal*	*High*
Standard	55	22	15
Night	50	25	18
Portrait	—	—	>300

Here's how to set your phone's camera to different resolutions:

1. **On the main menu, choose Camera (on Siemens SX1 choose Camera and then Snapshot).**

2. **After the Camera application launches, press Options on the left menu key.**

3. **Choose Settings (on Series 60 phones with integrated camera and camcorders, choose Settings and then choose Images).**

4. **Select Image Quality (the setting at the top of the setting screen).**

5. **Choose from Basic, Normal, or High.**

6. **Press OK on the left menu key.**

As long as you know how to transfer images to your PC (see Technique 52), you shouldn't worry too much about capturing images that result in larger file sizes. Just remember that there are limits to how big multimedia messages and e-mail messages can be — phones adhere to a standard to ensure service-provider networks can reasonably handle multimedia messaging. So, for picture or multimedia messaging, you may want to set the image resolution lower than the highest quality available.

Technique 16: Using the Timer Function

You know that feature on your digital camera that allows you to set a timer, run around to the front of the camera, and join your friends or family for a group picture? Did you know that many Series 60–based phones have that same feature? You can set your phone to snap a photo after a time delay of 10, 20, or 30 seconds.

Note   Older Series 60 models such as the Nokia 7650, Nokia 3600/20/50/60, and Siemens SX1 do not support the Self Timer feature.

How to stop the timer after it's activated

Imagine this scene. You set your phone's timer in order to snap a family group picture. As you're making your way around in front of the camera lens to join in the shot, your 2-year-old pulls off his diaper and proudly drops it into his grandmother's lap.

Yes, there are times when you need to cancel the timer's countdown on your phone's camera.

For all phones except the Sendo X, press the Cancel button on the right menu key or press the End key.

For the Sendo X, the Cancel or End buttons won't work. The countdown will continue even if you press Cancel. For this phone, you must close the camera application. To do this:

1. **With the camera launched and the timer counting down, press and hold the main menu button down.**

   You see the application shortcut menu.

2. **Highlight the Camera application, as shown in the figure.**

Highlight the Camera application in the application shortcut.

3. **Press the C (or Clear) key to close the Camera application.**

4. **Choose YES on the Exit Camera? screen.**

   Success! You've now stopped the timer on the Sendo X.

Tip Using the Normal profile, the self-timer will beep while counting down to snap the picture. And you'll hear the same artificial shutter click that you always hear when you capture an image. Most digital cameras let you silence the timer beep, and you can do the same on your camera phone. You just need to use the Silent profile (see your phone's user manual on how to switch profiles if you don't know how).

Tip How can you position your phone so that it will snap the picture without you holding it? Some phones, like Nokia's 7610, have flat bottoms and can stand on end. Others, like Nokia's 6600, you can lay on their side. Yes, your image will appear sideways on the display, but then you use the image rotate function (see Technique 14). Finally, there are some phones that you'll just have to prop up against a book or a notebook computer so that they're positioned properly for taking a picture.

To use the timer function to capture an image:

1. **On the main menu, choose Camera.**

2. **Select Options.**

3. **Choose Self-Timer (see Figure 3-6).**

Figure 3-6: Choose Self-Timer on the Options menu.

On the Sendo X, this immediately starts a 10-second countdown timer. The Sendo X does not let you choose different countdown times. For other Series 60 phones that support this feature, choose from the list of countdown times: 10 seconds, 20 seconds, or 30 seconds (see Figure 3-7).

Figure 3-7: On many Series 60 phones, choose a 10-second, 20-second, or 30-second timer (Sendo X features only a 10-second timer).

4. **Choose Activate by pressing the left menu key to start the timer (see Figure 3-8).**

Figure 3-8: Choose Activate to start your camera countdown timer.

Now position the camera phone the way you want and go get into the picture with your friends and family.

Technique 17: Replacing the Wallpaper of Your Phone with Your Own Images

Let's face it: One of the best features of smartphones is the ability to save your own images (or those you get from friends or family, or download from the Web) as the background of your phone's standby screen — also known as *wallpaper*.

For me, glancing down to check the time on my phone, viewing a text message, or just looking at an incoming caller's id, and seeing the face of one of my kids on my phone's screen gives me a little lift. It's one of those brief moments that make every day fun.

Yes, this is one of the few techniques in these pages that is also covered in most user manuals, but because it's so important, I'm using one of these 101 techniques to describe the steps — just in case you never opened the user manual. (Don't worry, I won't tell anyone.)

Here's how to save an image as your phone's wallpaper on newer Series 60 phones that include a Gallery application on the main menu:

1. **On the main menu, select Gallery.**

2. **Select a saved image.**

3. **Press Options on the left menu key.**

4. **Choose Set as Wallpaper.**

On older Series 60 phones (Nokia 7650, Nokia 3600/20/50/60, Siemens SX1, Nokia N-Gage, and Nokia N-Gage QD) you need to go to the Tools folder (Setup on SX1), choose Settings ➪ Phone ➪ Standby Mode ➪ Background Image (Wallpaper on SX1).

Because of the NOW! screen on the Sendo X, wallpaper is not supported on that phone.

You can also perform this same action on an image that you've just captured with your phone's camera.

Technique 18: Making Digital Photo Albums or Saving Images to Folders

You already knew your camera phone simplified your life by combining a digital camera and a phone — fewer things for you to carry around. Did you realize that your camera phone also features a digital photo album? That's right — forget the moldy family pictures in your wallet or the plastic accordion photo sleeves. Your phone features a customizable photo album that lets you organize photos, categorize them, even show them in mini slide shows.

Use your Gallery (called Photos on the Sendo X; Images on the Nokia 7650, Nokia 3600/20/50/60, Nokia N-Gage, and Nokia N-Gage QD; and Camera ➪ Images on the Siemens SX1) as your digital, portable photo album. In this section, I show you how to create a new folder, name it, move images into it, and display the pictures in a slide show.

Note   In the Siemens SX1, the digital photo album is one level deeper in the menus than the other Series 60 phones. On the SX1, open the Camera item and then the Images item to get to the image storage area.

To create a new photo album photo in your Gallery:

1. **On the main menu, choose Gallery Photos on the Sendo X; Images on the Nokia 7650, Nokia 3600/20/50/60, Nokia N-Gage, and Nokia N-Gage QD; or Camera ⇨ Images on the Siemens SX1.**

   If you have a Multimedia Memory Card (MMC) installed, you have to choose whether you want the new photo album to be stored in phone memory or on your memory card. It's an easy choice really: If you have a memory card installed and it's not full, you should store the new album there.

   There's no downside to choosing the memory card. Your access to the images stored there is just as quick as if you stored them in phone memory. Besides, because phone memory is shared by many applications on your phone—and the phone's operating system itself—you want to conserve phone memory when possible.

2. **To choose phone memory or memory card, simply move the joystick to the left or the right.**

   You'll see a small phone icon on the tab of the phone memory screen (see Figure 3-9) and the small icon of an MMC on the memory card tab (see Figure 3-10). Moving the cursor left or right moves you back and forth between the two tabs.

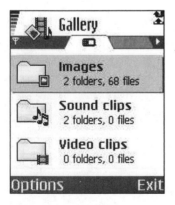

Figure 3-9: The small phone icon shows you the phone memory section of the Gallery.

Figure 3-10: The small MMC icon shows you the memory card content for the particular application. Always choose the memory card section if there's space available (see Technique 64).

3. **Select Options from the menu key.**

4. **Choose New Folder.**

5. **Enter a name for your new photo album (see Figure 3-11).**

   Be creative. These folders are the way to organize your images in your new digital photo gallery. You can create folders just for family photos and for summer vacations. Or you can categorize by topic — pets, travel, friends, and so on.

Figure 3-11: Be creative when naming folders in your photo gallery.

6. **Select OK on the left menu key.**

   The new folder you just created will appear in the list.

7. **To add some images to your new photo album folder, navigate to the images you want to move.**

   Remember: The images can be stored either in phone memory or memory card memory.

   If there is only one image, move the cursor to highlight the image in the list (see Figure 3-12). Do not open the image.

If you want to move more than one image, use Technique 28 to mark multiple items in a list (see Figure 3-13).

Figure 3-12: To move only one image to the new photo album folder, move the cursor to highlight the image.

Figure 3-13: To move several images at once, use Technique 28 to mark multiple items in a list.

8. **Choose Options.**

9. **Select Move to Folder (see Figure 3-14).**

Figure 3-14: Select Move to Folder to move images to a new location.

10. **On the Move To screen, highlight the memory area—phone memory or memory card—where your new photo album folder resides (see Figure 3-15).**

Note If you don't have a Multimedia Memory Card (MMC) installed, you'll only see Phone Memory listed on the Move To screen.

Figure 3-15: Pick the memory area where your new folder resides.

11. **Select Move (or center-press on the joystick).**

On the next Move To screen, you'll see a list of folders that exist in the memory area.

12. **Highlight the folder to which you want to move the image(s) (see Figure 3-16).**

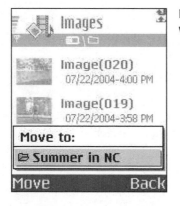

Figure 3-16: Pick the folder where you want to put the image(s).

13. **Select Move (or press down on the joystick).**

Now, you've stored images in your photo album folder. You can add or remove images from the folder whenever you want.

Here's the fun part. Now, instead of pulling out dirty, creased photographs from your wallet, show your friends a digital slide show of your pix. To display a mini slide show of images from a folder, follow these steps:

1. **On the main menu, navigate to the Gallery (Photos on the Sendo X; Images on the Nokia 7650, Nokia 3600/20/50/60, Nokia N-Gage, and Nokia N-Gage QD; Camera ⇨ Images on the Siemens SX1).**

2. **Choose Gallery (or Photos or Images).**

3. **Navigate to the folder containing the images you want to display.**

4. **Choose the folder.**

5. **Select the first image in the list of images contained in the folder.**

6. **Now, here's the trick, to move — in a slide-show manner — through the rest of the images in your folder, simply move the joystick to the right.**

   This action pages you through all the images contained in your photo album.

Simply hold your phone in front of a friend or a small group and you can easily show a series of family or vacation shots in a few seconds. Now, why did you think you needed to carry those pictures in your wallet, again?

Technique 19: Cannibalizing Messages for Their Images, Sound, and Video

Yes, you read that correctly — cannibalizing as in Hannibal the Cannibal, fava beans, and all that. This technique shows you how to strip and save the good parts of multimedia messages — like images, audio clips, and such — and throw away the rest.

Face it: You don't need old messages hogging up valuable space in your phone memory or MMC. But, you might make use of the cool pictures and sound bites that some of them contain.

This section shows you how to save message parts into your phone's gallery in an organized way, as well as how to delete the original message to save memory.

When you receive a multimedia message, it looks like it's all one document with text and images, perhaps even video or audio. In fact, under the covers, multimedia messages are really a collection of objects — text, image, audio, and video — all wrapped together in a format called Multimedia Message Service (MMS).

Here's how you can look at a multimedia message in object view:

1. **On the main menu, select Messaging.**
2. **Select Inbox (see Figure 3-17).**
3. **Select a multimedia message from your Inbox list (see Figure 3-18).**

Figure 3-17: Select your " Inbox.

Figure 3-18: Select a multimedia message (preferably one that includes text, an image, maybe even an audio attachment).

4. **Press Options on the left menu key.**
5. **Choose Objects (see Figure 3-19).**

   This is the key to this technique!

   At this point, you're in object view, and you'll see a list of all the objects that make up your multimedia message (see Figure 3-20).

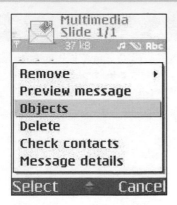

Figure 3-19: Choose Objects on the Options menu to switch to Object View.

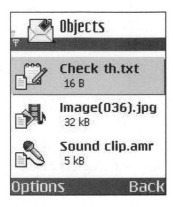

Figure 3-20: All the objects that make up your multimedia message.

6. **To save a particular object in the list, highlight the object.**

Note The text in a multimedia message is treated as an object, just like an audio, video, or image attachment.

Tip To delete an object in object view, simply highlight the object and then press the C (or clear) key. You'll see a confirmation message asking if you're sure you want delete the object. Press Yes on the left menu key.

7. **Press Options on the left menu key.**

8. **Choose Save.**

   Depending on the object type, your object will be saved in the default folder of its application (see the "Viewing objects in object view" sidebar). So a text object will be saved in the default folder for the Notes application, an image object in the Gallery, and so on.

Now, you can delete the message as you normally would if you want to free up some space in your phone memory or memory card. The valuable objects you saved will be stored away safely for later use.

Viewing objects in object view

In object view, you can view each multimedia object separately simply by selecting it. The proper native application for each multimedia type launches automatically and plays or displays the multimedia object you selected.

For example, when you select a text object in object view, the text appears in the Notes application (see the figure).

The Notes application shows
you the text objects.

An image object displays in the Image Viewer application (see the figure).

Images appear in the
Image Viewer.

An audio object plays in the Voice Recorder application (see the figure).

Audio plays in the Voice
Recorder.

And, you guessed it, a video object plays in the phone's video player, which in most cases is RealPlayer (see the figure).

Video plays in the video
player.

Technique 20: Uploading Images to Digital-Photo Services on the Web

Pictures are no fun unless you share them with someone. Showing people the pictures stored on your phone (or transferred to your PC via Technique 52) can be fun, but the people need to be close by.

What about your distant friends and relatives who live in other states or countries? How do you share images with them? Yes, you could use e-mail, but there is another way.

Remember the smart characteristic of your phone? It can connect to the Internet and remote computer servers in the same manner as a PC. Even better, several online services let you upload and store your images on the Web. Some of the services—like Kodak Mobile Service (also known as Ofoto) and DotPhoto—will print and send your images to your friends and family, wherever they are, for a fee. This type of service really expands what it means to share your smartphone pictures.

Here's how to access Kodak Mobile Service:

1. **Point your PC Web browser to** `www.kodakmobile.com`.

2. **Become a member.**

   Follow the instructions on the site. It's free. You sign up for the service separately.

3. **At this point, you can sign up for the service and you can upload images to the share areas of the site using your PC.**

   First, of course, you would need to transfer images from your phone to your PC using Technique 52. However, if you want to be able to access the site and upload images directly from your phone to the service, continue on with the following steps.

4. **Check the Web site to see if there is software (called *client software*) for your make/model of phone.**

   Follow the instructions on the Web site. The form will ask you for the name of your service provider and the make and model of your phone.

   If your phone and service provider are supported, you'll need to launch the browser on your phone.

5. **On the main menu of your phone, select Services (Web on Nokia 6260, Nokia 6620, Nokia 6630, Nokia 7610, and Panasonic X700; either WAP or Web on Sendo X; Internet on Siemens SX1).**

6. **Press Options on the left menu key.**

7. **Choose Navigation Options (see Figure 3-21).**

 Note   Navigation Options is only on Nokia 6260, Nokia 6620, Nokia 6630, Nokia 7610, Panasonic X700, and Samsung SGH D710. All other phones have Go to URL or Go to Address.

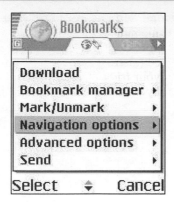

Figure 3-21: Choose Navigation Options on your phone's Web browser (only on Nokia 6260, Nokia 6620, Nokia 6630, Nokia 7610, Panasonic X700, and Samsung SGH D710 models).

8. **Choose Go to Web Address (see Figure 3-22).**

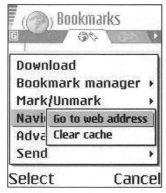

Figure 3-22: Choose Go to Web Address.

9. **In the address bar, type** www.kmobile.com **(see Figure 3-23).**

Figure 3-23: Enter the URL **www.kmobile.com**.

10. **Sign in to Kodak Mobile using your e-mail address or phone number and password that you just used to become a member (see Figure 3-24).**

Figure 3-24: Sign in to Kodak Mobile over your phone.

11. **Follow the instructions on the site to download and install the phone software.**

Signing in and setting up your phone for DotPhoto (www.dotphoto.com) is a very similar process. Check both of these sites and others that you find to see which is offering the best deal—usually a monthly charge for some number of prints per month. Uploading photos and sharing them with others over the Web is generally included in the monthly service fees or free altogether.

Technique 21: Adding Images to a Multimedia or E-Mail Message

If you've never used a Series 60–based smartphone before, I want to make sure to cover the basic technique of adding an image to a multimedia message or e-mail. Believe or not, this topic is generally given short shrift by phone user manuals.

In this technique, I show you the easy way to add an image to a multimedia message, some tips for customizing a multimedia message, and a neat way to preview what your message will look like to your recipient.

Here is a quick and easy way to snap a picture with your digital camera and add it to a multimedia message:

1. **On the main menu, select Camera.**

2. **Snap a picture by center-pressing the joystick.**

3. **Wait until the phone finishes saving the image, and then press Options on the left menu key.**

4. **Choose Send (see Figure 3-25).**

Figure 3-25: Choose Send on the Options menu.

5. **Choose Via Multimedia on the pop-up menu (see Figure 3-26).**

Figure 3-26: Choose Via Multimedia on the pop-up menu.

This action takes you into the multimedia message editor. Here you can add text and other multimedia objects.

6. **To preview what your message will look like, press Options.**

7. **Choose Preview Message (see Figure 3-27).**

   Now, you're in preview mode (see Figure 3-28).

8. **Press Back to return to the editor.**

   From there, you can send your multimedia message as you normally would.

Figure 3-27: Choose Preview Message on the Options menu to see what your message will look like to your recipient.

Figure 3-28: Preview mode for a multimedia message.

Moving the text in your multimedia message

Did you know you can put text at the beginning or the end of your multimedia message? If you follow the steps in Technique 21 — snap a picture and move it into the editor — the space for adding text will appear at the end of the message, below the image. Here's how to move that text to the beginning of the message:

1. **In the editor, press Options on the left menu key.**

2. **Choose Objects.**

This puts you into object view.

3. **Choose Options again on the left menu key.**

4. **Choose Place Text First (see the figure).**

Choose Place Text First to move your message text to the beginning of your multimedia message.

Now, your text has been moved to the top. You'll notice that change in preview mode and in the multimedia message editor. In object mode, the Option menu will now offer the choice Place Text Last, just in case you want to switch back.

Technique 22: Editing Images on Your Phone Using the PhotoRite Download

Some Series 60 phones, like Siemens SX1, feature image-editing software prein-stalled. This means when you turn the phone on — shebang! — you can edit your images as quickly as you snap them.

If your phone didn't come with image-editing software, don't panic. You can download cool image-editing software — like PhotoRite, the one I describe in this technique. It works in a similar way to the SX1's Image Fun application.

Image-editing applications on your phone allow you to add text, symbols, and graphics to your images. You can distort people's faces into funny alien-looking creatures, rotate objects, and change the sizes of objects.

PhotoRite also features an auto-enhance option that will sharpen up your images. You can also place themed frames around your images to create certain moods or celebrate events.

You can download PhotoRite from Zensis (www.zensis.com). Like all the other shareware mentioned in these pages, you should try it before you buy it. Zensis offers a 15-day free trial of the software, which you can download and install (by reading Technique 72 on installing Symbian software).

Here is a example of using PhotoRite to enhance a photo, customize the image with an effect, and then add a frame:

1. **On the main menu, choose PhotoRite (see Figure 3-29).**

Figure 3-29: Launch PhotoRite.

You can browse to an image in phone memory or in a memory card, or you can capture a new image inside PhotoRite. In this example, I show you how to capture a new image.

2. **Press the joystick to the right until you see a camera icon on the tab at the top of the screen.**

3. **Center-press the joystick to capture a new image.**

4. **Press Options on the left menu key.**

5. **Choose PhotoRite Auto-Fix (see Figure 3-30).**

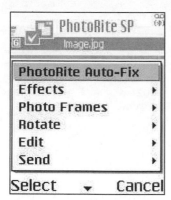

Figure 3-30: PhotoRite's auto-enhance feature.

Now, we'll add an effect.

6. **Press Options again.**

7. **Choose Effects (see Figure 3-31).**

8. **Choose Slim Up! (see Figure 3-32) from the pop-up menu.**

Figure 3-31: Add an effect.

Figure 3-32: Choose the
Slim Up! effect.

9. **Choose Extra Slim Up! from the pop-up menu.**

10. **Press the joystick up to expand the image or down to squash the
    image.**

11. **Press Options on the left menu key again.**

12. **Choose Photo Frames (see Figure 3-33).**

13. **Choose Happy Birthday (see Figure 3-34).**

    Now, your photo is customized (see Figure 3-35) and is ready to send
    or transfer to your PC.

Figure 3-33: Choose
Photo Frames.

Figure 3-34: Choose the Happy Birthday frame.

Figure 3-35: Your customized photo.

Buying and Using Ring Tones

◆   ◆   ◆   ◆

In This Chapter

Finding out how to make ring tones with your voice-recorder application

Discovering how to make your favorite song a ring tone

Making your own ring tones with a special shareware application

Finding the best sources for ring tones on the Web

◆   ◆   ◆   ◆

Analysts say that, when all the dollars are counted, the people of the world will likely have spent over $4 billion in 2004 on ring tones. Let me repeat that. $4 billion! That's like Oprah Winfrey and Donald Trump money. Ring tones are big business by any measure.

Why do people spend all that money to alter the tune their cellphones play while ringing? It's anyone's guess, really. But *my* guess is that it's all about individuality and self-expression. Like clothing, hairstyles, tattoos, and body piercings, ring tones are another way you can express yourself, show your own style, and communicate to others things they won't find on your résumé.

Fortunately, for you, your Series 60 phone excels in this category of customization (just as it excels in almost every other category — do you see how amazing these phones are yet?). In this chapter, I clue you in to the best places to find, purchase, and download ring tones; how to make your own ring tones without purchasing anything; and how to make your own ring tones using a shareware application called Ringtone Studio.

Technique 23: Buying and Downloading Ring Tones

Many sources for ring tones are out there. The best places include your phone manufacturer and your service provider — you have easy access to both of these sources if you followed the steps in Technique 4.

Your phone manufacturer and service provider each offer two ways to purchase and download ring tones — via browsing their Web sites using a PC or
accessing their Web sites directly through your smartphone.

Here are the basic steps for purchasing and downloading ring tones using
your PC:

1. **Point your Web browser to the desired Web site.**

   Here's a list of some of the most common manufacturers and service
   providers and their Web sites:

   - **AT&T Wireless:** www.attwireless.com
   - **Cingular:** www.cingular.com
   - **LG:** www.lge.com
   - **Lenovo:** www.lenovo.com
   - **Nokia:** www.nokia.com
   - **Orange:** www.orange.com
   - **Panasonic:** www.panasonic.com
   - **Samsung:** www.samsung.com
   - **Sendo:** www.sendo.com
   - **Siemens:** www.siemens-mobile.com
   - **T-Mobile:** www.t-mobile.com
   - **Verizon Wireless:** www.verizonwireless.com
   - **Vodafone:** www.vodafone.com

2. **Log onto the site if you've previously registered.**

3. **Identify which phone you're using, if that information isn't already
   kept in your user profile at the site.**

4. **Locate the section of the site that offers ring tones.**

   Most sites let you listen to samples of the ring tones.

5. **Make your choice and transact the purchase.**

   If the site is your service provider's, you'll likely be given the choice
   of providing a credit-card number or putting the charge on your next
   monthly bill.

6. **Provide your smartphone's number (if it's not already stored in
   your user profile at the site).**

   The ring tone will be sent to your phone via a special configuration
   SMS message. (See Technique 2 for more information on configuration SMS messages.)

7. **Save the ring tone to your phone's memory.**

8. **Assign the new ring tone to one of your profiles (see Figure 4-1).**

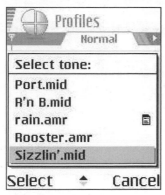

Figure 4-1: Assign your new ring tone to a profile.

Now, you've finished! Perform these steps just once and you'll get the hang of it. It'll go much faster the second time through.

Here's how you purchase and download a ring tone by accessing a Web site via your phone:

1. **Point your phone browser to the Web site (see Figure 4-2).**

   You'll likely have bookmarks already set for both your phone maker and your service provider, if you don't access the sites via your PC Web browser and have bookmarks of the sites sent to your phone (most phone makers and service providers offer such bookmark capability now).

Figure 4-2: A ring tone site bookmark on mMode's Web site as viewed through a phone browser.

2. **Log onto the site if you've previously registered.**

   Many service-provider Web sites identify your phone automatically when you connect via your phone.

3. **Identify which phone you're using, if that information isn't already kept in your user profile at the site.**

4. **Locate the section of the site that offers ring tones.**

   Most sites let you listen to samples of the ring tones.

5. **Make your choice and transact the purchase.**

   If the site is your service provider's, you'll likely be given the choice of providing a credit-card number or putting the charge on your next monthly bill. Because you're browsing via your phone, the site may allow you to transact the purchase using your phone's e-wallet.

6. **Provide your smartphone's number (if it's not already stored in your user profile at the site).**

   The ring tone will be either sent to your phone via an MMS message, or it will be downloaded to your phone directly from the site.

7. **Save the ring tone to your phone's memory.**

8. **Assign the new ring tone to one of your profiles.**

As with a PC Web browser, the more you use your phone's browser the more comfortable you'll feel with it. Purchase a ring tone once or twice following these steps, and you'll enjoy the convenience of buying online without being chained to a PC.

You may also want to try the following sites to download ring tones:

✦ **Handango:** www.handango.com

✦ **ToneGuys:** www.toneguys.com

✦ **Zedge:** www.zedge.no

✦ **Zingy:** www.zingy.com

Technique 24: Creating Your Own Ring Tones

As I mention in the chapter introduction, the ring-tone business is booming. But if you'd rather not contribute to the huge pool of money spent on ring tones, you're in luck. In this technique, I show you how you can create your own ring tones.

Many for-sale ring tones feature popular artists playing their hit tunes, but if you've already purchased these songs, you can use the digital versions of them as your ring tone for free. As long as this is for personal use and you've paid for the music—either by buying a CD from a music store or from down-loading the tune from a digital music service—there is nothing illegal about it.

Note Many Series 60 phones support MP3 and AAC digital music formats for use as ring tones.

Here's how to make your favorite digital tune into a ring tone:

1. **Get the tune into digital form on your PC.**

   If, for example, your tune is on an audio CD, you need to convert or rip the tune into one of the Series 60 supported formats (AMR, WAV, as well as MP3 and AAC in many models). You'll find many PC applications that rip CDs. Windows Media Player, which comes integrated into Windows, for example, will do the job.

2. **Locate the tune on your PC hard drive.**

   If you know the name of the tune or the format it's in (like MP3, AAC, or WAV), you can search for it using Windows Explorer (see Figure 4-3).

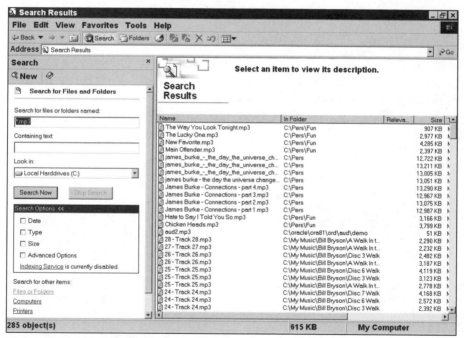

Figure 4-3: Search for tunes using Windows Explorer.

3. **Use your phone's PC suite software (see Technique 52) to convert the file (if applicable).**

4. **Use your phone's PC suite software to transfer the tune to your PC (see Figure 4-4).**

5. **Locate the tune on your phone (see Figure 4-5) using the File Manager (see Technique 31).**

6. **Select the tune by center-pressing on the joystick.**

This will launch the Music Player application (see Figure 4-6).

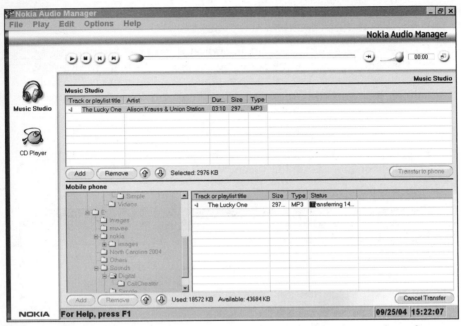

Figure 4-4: Transfer the tune to your phone using your phone's PC suite software (in this example, Nokia's Audio Manager).

Figure 4-5: Locate the tune on your phone using the File Manager application.

Figure 4-6: Open the tune
to launch the Music Player.

7. **Press Options on the left menu key.**

8. **Choose Set as Ringing Tone (see Figure 4-7).**

Figure 4-7: Choose Set as
Ringing Tone.

You'll see a message confirming that the normal profile ring tone has been changed.

Another cool way to create your own ring tone is to use your smartphone's voice recorder. The secret: The integrated voice-recorder application on your phone records in AMR format. This is a format supported by your phone for ring tones.

To me, some of the funniest and most distinctive ring tones aren't musical at all; they're just voice based. Here are some examples:

✦ For those who never know whose phone is ringing: "Dean's phone. Dean's phone ringing. Dean, please answer your phone."

✦ For sports fans: "Any team can have a bad century, but now the Red Sox are World Champions! Yes! Go Red Sox!"

✦ For parents who just can't get enough of their little ones: A recording of their children singing.

✦ For TV fans: A recording of a few seconds of a favorite TV show.

Follow these steps to record and set a ring tone using your phone's voice recorder:

1. **On the phone's main menu, select Extras.**

2. **Select Recorder (see Figure 4-8).**

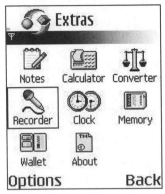

Figure 4-8: Launch the Voice Recorder.

3. **Select the red button to start recording.**

4. **Say your message loud and clearly into the microphone (usually positioned just below the keypad).**

Are you ready for video ring tones?

This should be one of those techniques that comes with a warning label: Don't try this on a wimpy non–Series 60 phone! A company called Psiloc (www.psiloc.com) created an application called Vision that lets you create video ring tones.

Basically, when someone calls you, your phone plays a video with sound instead of just a traditional ring tone. Vision offers full-screen mode so the video is very visible, and you can assign different videos to different contacts, so you can use the application like you use thumbnail contact images—to see rather than just hear a unique tune for your special frequent callers.

You can download Vision from Psiloc's Web site and try it before you buy it. As of this writing, Vision costs 14.95€ ($18.75). But, remember, this works on your smartphone because of Series 60/Symbian multitasking capability, so it won't run on your average smartphone or camera phone. Psiloc lists compatible phones on its Web site.

5. **Press Stop on the right menu key to stop the recording (see Figure 4-9).**

Figure 4-9: Press Stop to end the recording.

6. **Press the Play button to listen to your sound clip.**

   Rerecord it as necessary.

7. **Press Options on the left menu key.**

8. **Choose Rename Sound Clip (see Figure 4-10).**

Figure 4-10: Give your ring tone a name by choosing Rename Sound Clip.

9. **Enter a new name in the text entry field.**

10. **Press OK on the left menu key.**

    Now, you can leave the voice recorder and go customize a profile.

11. **On the main menu, select Profiles.**

12. **Select the Normal profile or some other profile you want to change.**

13. **On the pop-up menu choose Customize.**

14. **Select Ringing Tone.**

    In the list of ring tones, you'll see the name of the ring tone you just recorded (see Figure 4-11).

Figure 4-11: You'll see the name of the ring tone you just recorded in the list of ring tones.

15. **Select your new ring tone.**

    Now, you'll hear your sound clip when people call.

Tip   You may need to increase the volume of a recorded ring tone in your Profile setting. Generally, voice recordings sound softer than musical ring tones, and you may not hear it very clearly if you don't adjust the volume.

Finally, there are several ring-tone editors that you can use to customize existing ring tones or create new ones. One good one is called Ringtone Studio. You can download it from WildPalm (www.wildpalm.com). Install it by following the steps in Technique 72.

Ringtone Studio supports MIDI (.mid) format ring tones. Many of the ring tones present on your smartphone are in this format. With Ringtone Studio you can modify the tunes, change the speed, or shorten the tune. You can even remove or modify individual instruments that make up the tune.

Here are the general steps for using Ringtone Studio:

1. **On the main menu, select Ringtone Studio (see Figure 4-12).**

2. **Inside Ringtone Studio, press Options on the left menu key.**

3. **Choose Open.**

    Now you'll see a list of MIDI format ring tones that are present on your phone (see Figure 4-13).

Figure 4-12: Launch Ringtone Studio.

Figure 4-13: Ringtone Studio shows you the list of MIDI ring tones on your phone.

4. **Select one of the ring tones.**

   At this point, you'll see a list of instruments used to play the notes in this tune (all this information is stored digitally within a MIDI file).

5. **Choose Options ⇨ Edit Track to change the volume, mute, and add effects to individual tracks (see Figure 4-14).**

6. **Similarly, choose Options ⇨ Edit Song to change the speed, remove moments of silence, and select different start and end points of the tune (see Figure 4-15).**

Figure 4-14: By choosing Options ⇨ Edit Tracks, you'll see this list of modifications you can make to a track.

Figure 4-15: Choose Options ⇨ Edit Song to adjust the speed and adjust the start and end points of the tune.

7. **When you're satisfied with your changes, choose Options ⇨ Save to store them back in the MIDI file.**

Back in Profiles, you can assign your new ring tone to a profile.

Tip

Remember to try shareware applications like Ringtone Studio before you buy them.

Increasing Your Productivity

Your old cellphone made you more productive simply because you could stay in touch while away from your desk or home. Your new Series 60–based smartphone can make you more productive in many more ways. Although service providers are marketing smartphones simply as amusing picture-taking toys, these phones have powerful capabilities that can help you stay on schedule, be more productive, and carry your most important information with you wherever you go.

This chapter shows you some shortcut techniques that you won't find in any user manual. You'll be able to launch applications faster, switch between different applications, even use an undocumented copy and paste feature that will save you a tremendous amount of time as you enter new contacts, browse the Web, and more. You'll also find out how to make your phone work like one of those popular USB key-ring storage devices and a sophisticated appointment keeper.

Sure, PDAs and laptop computers have their place. But when you only want to carry one device, you may find your Series 60–based smartphone phone is the only one you really need. This chapter shows how to get more than some fun snapshots from your camera phone.

Technique 25: Launching Applications by Number

You can find out about speed-dialing techniques in Chapter 2, but here I show you how to speed-launch your applications. That's right — speed-launching applications. Sounds strange, doesn't it?

You'd be hard pressed to find this technique in your phone's user manual. Like many of the other techniques in this book, the developers of the Series 60 User Interface have made shortcuts like this available to end users, though until now, end users weren't formally informed of them outside of word of mouth.

 Note   This technique does not work on the Siemens SX1.

There is no need for you to assign a speed number to your applications as you do when you assign a number to a contact's phone number for speed-dialing. Your phone automatically assigns numbers to the first nine applications based on their position in the menu.

The application in the upper-left-hand corner of your main menu is assigned number 1, the application next to it to the right in the same row is assigned number 2, and so on. In Figure 5-1, Gallery = 1, Messaging = 2, Contacts = 3, Calendar = 4, Camera = 5, Video Recorder = 6, RealOne Media Player = 7, Services = 8, and Application Manager = 9.

Figure 5-1: Applications are automatically assigned speed-launch numbers based on their position.

The applications in your main menu are automatically assigned numbers that you can use to launch them with the press of one key. Good luck finding this technique in your user manual!

 Tip   Only the first nine applications are assigned numbers in your main menu *even* if you have more than nine applications listed there (and you probably do). So how can you speed-dial the applications that are positioned somewhere after the ninth one? There's really only one way: You have to move the application up in the main menu so that it gets assigned one of the 1 through 9 slots. Therefore, you should place your top nine most frequently used applications in the top nine spots. To find out how to move applications and folders in the main menu, see Technique 37.

Tip

Hey, wait a minute! I'm talking only about the numbers 1 through 9 for launching applications, but there's also a 0 key on your phone. What's up? Well, in the Main menu, pressing the 0 key does nothing, But, in standby mode, pressing and holding the 0 key automatically launches the Web browser (also called Services on some Series 60 phones). Unlike the speed-launch keys described in this section, you have to hold the 0 key for a couple seconds before it activates the browser.

To launch an application by number:

1. **Press the main menu key to go to your main menu.**

2. **Choose the application you want to launch, check its position, and press the appropriate number key.**

   You'll see the cursor jump to highlight the application you've chosen, and then the application will be launched automatically.

Technique 26: Copying and Pasting Data

I'm sure you value the productivity boost you get from copying and pasting on your PC or iMac. On desktop computers, not only can you copy and paste information within a single application — like Microsoft Word — but you can also copy and paste between applications — like between a word processing application and a spreadsheet application.

On a desktop computer, this feature saves you time because you don't have to retype anything — names, text, data, whatever. You just type it once and then copy it and paste it wherever you need it, even if it's another document or application.

Guess what? Your Series 60–based smartphone offers the same capability. But, no, you won't find it described in your phone's user manual. Whew! Good thing you bought this book.

On your phone, you can copy and paste any information that you can enter or edit. Here's an example using the contact database. To copy and paste information:

1. **On your main menu, choose Contacts.**

2. **Select Options.**

3. **Choose New Contact.**

4. **Enter a first name and a last name of a new contact.**

5. **Scroll down to Telephone.**

6. **Enter a new telephone number (see Figure 5-2).**

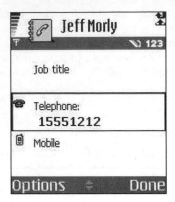

Figure 5-2: Enter a new telephone number with country code (+1 in the U.S.) and the area code.

Now, here's the trick: When you enter multiple numbers for a new contact, all the person's phone numbers will probably be in the same area code. Sometimes multiple numbers for a single individual only vary by a single digit.

Thus, it would really increase your productivity if you didn't have to reenter parts of the number several times. Well, with cut and paste, you don't.

7. **To copy the number, press and hold the Edit key.**

8. **While holding down the Edit key, move the cursor key to the left (or to the right if the blinking text cursor started at the beginning of the number).**

With each click of the joystick, you'll see another digit of the number become highlighted (see Figure 5-3).

Figure 5-3: Press and hold the Edit key while moving the joystick to highlight data.

After highlighting the whole number, you want to copy it.

9. **Still keeping the Edit key pressed down, choose Copy from the left menu key.**

   There will be no change on the screen. Just like on your PC, the information is copied and stored under the covers. You won't see anything until you go to paste the data somewhere.

Tip   Holding down the Edit key toggles Copy and Paste on your phone's menu keys. If you stop pressing the Edit key, your menu keys return to their usual Contact application choices. If you press and hold the Edit key, it changes the menu keys to Copy and Paste. Remember, though, that you need to have something highlighted in order to see the Copy and Paste choices; otherwise, the menu keys will just appear blank while you hold down the Edit key.

Tip   You can also press the Edit key twice quickly—when you have data highlighted—to bring up the Cut, Copy, Paste menu (see Figure 5-4). This is an alternative to holding down the Edit key.

Figure 5-4: Press the Edit key twice quickly to display the Cut, Copy, Paste menu.

10. **Scroll down to the mobile number of this contact record.**

11. **Press and hold the Edit key.**

    Paste will appear on the right menu key.

12. **Choose Paste.**

    The data you copied from Telephone will also appear in Mobile.

Remember: Just as on your PC, this technique works across applications on your phone. This section's example describes cutting and pasting a number within Contacts. But, using the same technique, you can copy a name from Contacts into your Calendar—when you have to schedule an appointment with one of your colleagues. Or, you can copy and paste information from a note to a message.

Tip   Practice the technique a few times so that you can do it quickly without thinking. Then you'll be able to save valuable time by never having to reenter information on your phone.

Technique 27: Using the Application Shortcut Menu to Switch between Open Applications (including Closing Apps with C)

You're probably familiar with the word *multitasking* (the ability to pause in the middle of a task, do something else, and then pick up the original task where you left off). From working with PCs, you know that you can switch between applications like a word processor and a spreadsheet effortlessly. But what about your Series 60–based smartphone?

An undocumented special multitasking menu on your smartphone can provide a quick speedup in the way you get things done with your phone. You won't find these instructions in your user manual, so pat yourself on the back for buying this book!

The special menu is officially called the Application Shortcut menu. In the following example, you first launch Contacts and then the Calendar. Then you use the Application Shortcut menu to switch between these applications.

To switch between applications using the Application Shortcut menu:

1. **On the Main Menu, choose Contacts.**

2. **Search on a particular letter (like *b*) just to bring up a list of contacts (see Figure 5-5).**

   Do this to prove that the Application Shortcut menu pauses an application and restarts it from the same point when you return.

Figure 5-5: Search in your Contacts list.

3. **Press the main menu key.**

4. **Choose Calendar.**

5. **Move the joystick to the right to display the next Saturday.**

   Again this is to prove the point that pausing an application works properly.

6. **Press the main menu key again.**

7. **Press and hold the main menu key.**

   This action displays the Application Shortcut menu (see Figure 5-6). You should see three boxes on the Application Shortcut menu — Calendar, Contacts, and Telephone.

Figure 5-6: The Application Shortcut menu allows you to switch between open applications on your phone.

 Note

On Series 60 phones, the Telephone application is always running. So, wherever you activate the Application Shortcut menu, you'll see the telephone application (see Figure 5-7).

Figure 5-7: The telephone application is always running, so you'll always see it on the Application Shortcut menu.

Note The Application Shortcut menu displays at most three applications at one time, even if more applications are active. The way to see the other open applications is to scroll up or down using the joystick while the Application Shortcut menu is visible.

Tip How can you tell if the menu is actually larger than the three applications you see when the Application Shortcut menu first becomes visible? Look for the up/down arrows in the center of the task bar at the bottom of the screen (see Figure 5-8). These arrows show you that scrolling will show you more choices.

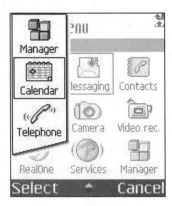

Figure 5-8: Look for the up/down arrows in the center of the task bar at the bottom of the screen. They show you that the menu has more choices than the three you see. You can see the additional choices by scrolling up and down.

8. **To switch to one of the applications, highlight it with the cursor and select it (choose Select off the menu key or center-press on the joystick).**

   This will display the chosen application exactly at the point where you last left it. Thus, if from our example you select Contacts, you'll see the results of the last search you performed (in this example, the list of your contacts starting with the letter *b*). If you select Calendar, this shows your schedule for the upcoming Saturday.

Use this powerful Application Shortcut menu to get stuff done faster! If you're interrupted while creating a new SMS message, simply press the main menu key to jump to something else, and then pick up where you left off later by choosing Messaging from the Application Shortcut menu.

Closing running applications

What's a quick way to close applications that are running on your phone? Use the special C (or clear) key while in the Application Shortcut menu. Here's how:

1. **Bring up the Application Shortcut menu by holding down the main menu key for a few seconds.**

2. **Highlight an application in the Application Shortcut menu by moving the cursor up or down.**

3. **Press the C (or Clear) key.**

    This displays a confirmation window (see the figure).

Your phone will display a confirmation screen asking if you really want to exit the application.

4. **Press Yes to close the application.**

    This returns you to the Application Shortcut menu. The Application Shortcut menu no longer shows the application you just closed.

Why should you close running applications? Even though the Symbian OS has an exceptional power-management subsystem, every launched application takes up some of your phone's memory, even if it's in a paused state. So your phone will actually run things faster and more efficiently if you close unneeded applications.

Technique 28: Selecting Multiple Items for Deleting, Moving, and More

This technique fundamentally changed the way I use my phone. Before I learned this technique, I deleted SMS and MMS messages from my inbox, deleted contact records, and moved images *one at a time*. If I count up all the seconds I wasted doing these single actions instead of marking several items and acting on them all simultaneously, I could have written a book in that time!

That's right—you can easily mark items in a list for deleting or moving. Did you know that? Your user manual certainly never told you about this feature. You either had to stumble across it yourself or hear it by word of mouth as I did from a phone guru (thanks, Brian Woods!).

If you're like me, you'll end up using this technique a lot—for cleaning out incoming messages from your inbox, for organizing your images in folders, for copying contact records from your SIM phonebook to your phone's contacts directory.

Here's a simple example—deleting multiple SMS messages from your inbox:

1. **On the main menu, select Messages.**

2. **Choose Inbox.**

   Let's assume you have more than one SMS message in your inbox. And, that your cursor is at the top of the list, as shown in Figure 5-9.

Figure 5-9: Start with the cursor at the top of your inbox's list of messages.

3. **Hold down the Edit key, and then move the joystick down.**

   At the right end of the messages, you'll see checkmarks next to each message you've highlighted (see Figure 5-10). This key combination marks a set of messages upon which you'll perform some action.

Figure 5-10: Notice the checkmarks to the right of each message. This tells you the messages are marked and ready for you to act on them.

4. **Press the C (or Clear) key to delete the messages.**

   The phone displays a confirmation message asking if you really want to delete the messages (see Figure 5-11).

Figure 5-11: A confirmation message asks if you really want to delete the marked messages.

 As with marking, you can also use the menus to delete. With the messages marked, choose Options. Then select Delete (see Figure 5-12).

Figure 5-12: The Options menu features a delete choice that you can use as a slower alternative to the C key for deleting messages or files.

5. **Choose Yes to delete the set of marked messages from your inbox.**

Remember: Deleting a set of messages is just one example of using this powerful productivity-enhancing technique. It's also very handy for moving a number of messages into a new folder or copying selected contacts from your SIM card's phone directory to your Contacts directory—without copying the entire list.

Technique 29: Marking and Unmarking Items Using the Options Menu

You can also mark and unmark messages, images, or files using the Options menu. In your inbox:

1. **Choose Options.**

   The Options menu features a Mark/Unmark choice.

2. **Select Mark/Unmark (see Figure 5-13).**

Figure 5-13: You can mark an individual message, mark all the messages in a folder, or unmark all the messages in a folder using the choices on the Mark/Unmark menu.

3. Choose Mark to mark an individual message (see Figure 5-14) or choose Mark All to mark all the messages in your folder (see Figure 5-15).

Figure 5-14: Choose Mark to mark one message in your folder for deleting or moving.

Figure 5-15: Choose Mark All to mark all the messages in your folder for deleting or moving.

Unfortunately, using the menus is not as fast as using the key combination (Edit key/joystick) described in this technique. Moreover, using the key combination, you can quickly mark several messages but not all the messages within a folder, making the key combination more powerful and flexible.

Technique 30: Unmarking Multiple Items Using the Edit Key

After you've marked a set of messages, images, or files, you can quickly edit it by unmarking a few files you didn't intend to include using a very similar key combination. To unmark selected files within a list of marked files:

1. **Use the joystick to scroll to a marked file (which has a checkmark to the right of it).**

2. **While holding down the Edit key, press down on the joystick.**

The check disappears (see Figure 5-16).

Figure 5-16: While holding down the Edit key, press down on the joystick to unmark a file.

You should also note that if you now move the joystick up or down, you can unmark a whole group of previously marked files. The Edit key/joystick actually toggles between marking and unmarking a list of messages, images, or files.

Technique 31: Finding Files and Multimedia on Your Phone

Did you know that nearly all Series 60–based phones (except the Nokia 7650) have a File Manager? The integrated Series 60 File Manager is very similar to the Windows Explorer you know on your PC. As on your PC, you can use the Series 60 File Manager to create new folders, move and copy files, and search for files.

The File Manager on your phone makes it easy to organize your information and data. Plus, it comes in very handy when you've lost track of something.

Think it's impossible to lose files on your phone? I personally have misplaced almost every kind of information or data on my phone at one time or another, but images are the items I lose most often. Fortunately, I've learned how to quickly search for items on my phone based on whatever information I remember about the file.

1. **To find a file on your smartphone, on the main menu, select Tools (Extras on the Siemens SX1, DocView on Sendo X).**

2. **Select File Manager (see Figure 5-17).**

Figure 5-17: The File Manager lives inside the Tools folder on most Series 60 phones, but on the Siemens SX1 it resides in the Extras folder, and on the Sendo X it's the DocView application on the main menu.

3. **Press Options.**

4. **Choose Find (Search ⇨ Filter on the Siemens SX1), as shown in Figure 5-18.**

Figure 5-18: Choose Find to display the File Manager's Search box.

5. **Choose either Phone Memory or Memory Card.**

Unfortunately, you cannot search across memory areas.

Tip

If you really have no idea where to find a file, search both areas one right after another. Simply follow these steps to search in Phone Memory, and then start again at Step 4 (you won't need to launch the File Manager a second time) and choose Memory Card for your next search.

6. **In the Search box, enter whatever information you remember about the missing file.**

If you remember the full name, go ahead and type it. If you remember only part of the name, that will work as well. Even if you only remember the type of the file—like JPG—just enter that, including the period (see Figure 5-19).

Figure 5-19: Even if you enter only part of the file's name, the File Manager will find it.

Tip

Press the * key on your phone to display a submenu of punctuation (period, question mark, exclamation point, and so on) and symbols (parentheses, brackets, @ signs, and so on), as shown in Figure 5-20.

Figure 5-20: Press * to display a submenu of punctuation and symbols.

Note

File Manager searches are case-sensitive, meaning it matters whether you enter a capital letter or a lowercase letter. In this example, if I had entered ".Jpg," the search would return no matches. Because I entered ".jpg," it returns many matches.

To switch between upper- and lowercase for entering text, press the # key on your phone. This toggles text entry between upper- and lowercase. An indicator appears in the top-right corner of the screen or menu whenever text is being entered.

The File Manager application searches for full or partial matches every time, so if you've entered only a small amount of a missing file's name, the application will display every file it finds that has characters you entered anywhere in the file's name (see Figure 5-21).

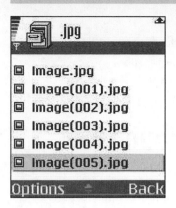

Figure 5-21: The search results show every file that matches your search string.

Technique 32: Creating Your Own Menu of Shortcuts Using the Go To List

The Series 60's Go To menu lets you create your very own menu of shortcuts, excluding anything you don't consider "top priority." It's a great way to speed up your path to all your favorite applications.

Note On the Siemens SX1, the Go To menu is called Favorites. And, by default, it resides inside the Extras folder. The Sendo X does not have a Go To menu

Keep in mind that only certain preinstalled applications and functions — like Notes, Inbox, Calendar, and so on — can be placed on the Go To menu. Still, the Go To menu provides enough flexibility to allow you to work faster and smarter. You'll know which applications and functions are candidates because their Option menus will feature an Add to Go To item.

Note Don't worry about how moving around items in your main menu or moving applications from folder to folder will affect your Go To menu. They won't because the Go To menu's shortcuts are updated automatically. So, the Go To menu will work no matter how you move things around.

Tip If you use your Go To menu frequently, you'll want to make sure you can access the Go To menu with just one key-press. Do this by assigning the Go To menu (again Favorites on the Siemens SX1) to the left or right menu key in your phone's standby mode. Read Technique 13 for instructions on assigning applications to your left and right menu keys.

In the following example, you add the Note function to your Go To menu. To add an application or function to your Go To menu:

1. **Navigate to the application or function that you want to add.**

   In this example, you'll navigate to the Notes application. To begin, press the main menu key.

2. **Select Notes (Organizer on the Siemens SX1).**

3. **Press Options.**

4. **Choose Add to Go To (Add to Favorites in the Siemens SX1).**

   This action displays the message Shortcut added to Go To (Favorites on Siemens SX1).

Technique 33: Using Your Phone as a Temporary Storage Device

Thinking about buying one of those USB key-ring-style storage devices? They're great for transporting documents from work to home or backing up individual files. You can move files back and forth from a PC to the storage device by plugging the penlight-sized devices into a USB port on your PC. And you can conveniently carry it around in your pocket or attach it to your key chain.

But wait! You may not need to spend the money. Your phone can act as a temporary storage that's just as convenient and portable. With an inexpensive memory add-on, you can a store large number of PC files on your phone, in addition to images, multimedia, and applications. You can connect your phone to your PC in a number of ways (see Chapter 9), both wired and wireless. And, you can carry your phone comfortably in a belt-attached case or in your pocket or purse. Besides, you probably already carry your phone everywhere—why carry yet another digital device if it's not necessary?

Many phones now come with an extensible memory card right in the box. Memory cards come in many formats, are made by a variety of manufacturers, and currently range in size from 8MB to 512MB. All Series 60–based phones support the MultiMediaCard (MMC) format, and several support additional card formats like SecureDigital (SD). The Series 60–based Nokia 6600, for example, comes standard with either a 16MB or 32MB MMC, depending on the sales package.

Don't let that stop you from upgrading the memory in your phone. Memory is relatively cheap, so purchase the largest card that your phone supports. Memory cards range in price from under $45 up to about $250. Consult your phone's manual to verify the types of cards that your phone supports.

Be careful when inserting or removing memory cards. These small cards can be damaged if inserted or handled improperly. Read your phone's user manual for instructions on installing, formatting, and removing phone memory.

Most Series 60–based phones from Nokia dock the memory card under the phone battery, next to the SIM card. Other models, such as the Nokia N-Gage QD, Panasonic X700, Sendo X, and Siemens SX1, have the memory card slot on the exterior of the phone.

I frequently use my smartphone to transport documents from my work PC to my home PC for "after-hours" editing. Just before I leave the office, I connect my phone to my PC using the infrared port, and then I use the PC suite software to quickly move my work-in-progress files to the memory card of my phone. Once at home, I connect the phone to my PC and send the files over. I'm traveling light but still productive in my spare moments at home.

For information on using your PC suite software see Technique 49. To stay flexible, install your PC suite software on both your work and home PCs. That way you can "courier" files back and forth on your phone anytime you like.

Technique 34: Accessing Your Phone's Applications While You're Taking a Call

Ever need to look up someone's contact info on your phone while you're taking a call? Have you ever been on the phone with someone and felt like sending a quick text message to alert a friend or colleague of a change in plans? Have you ever been playing a game on your phone and been frustrated that you had to start over after an incoming phone call closed the game application? No worries. Pat yourself on the back again for buying a Series 60 phone because it can multitask.

That's right. You can still use all your phone's functionality (with the exception of some data services, like browsing and e-mail) while you're on the phone. That's because the multitasking capability allows multiple applications to run (and interact) simultaneously. So, you can do all of these without issue, including getting back to setting the high score in that cool game you just downloaded!

Of course it helps if you can see what you're doing while you work with your phone. And, that's difficult if you're holding the phone up to your ear to engage in conversation.

This technique works best if you use a headset or use your phone in hands-free mode. (Read Technique 95 for instructions on attaching your headset.) To put your phone in hands-free or speakerphone mode, simply press the right menu key which, during a phone call, shows "Loudsp" (for Loudspeaker), as shown in Figure 5-22. Some Series 60 phones show "Handsfree".

Figure 5-22: Just press Loudspeaker (Handsfree on some Series 60 phones) during a phone call to switch to speakerphone mode, where you can hold the phone away from your ear and still hear and talk to the person on the other end of the line.

Alternatively, you can open up the Options menu on the left menu key during a call and choose Activate Loudspeaker from the list (see Figure 5-23).

Figure 5-23: You can also choose Options ➪ Activate Loudspeaker to go in hands-free or speakerphone mode.

Tip

Remember, there's no magic here. If you press End Call on the menu key or the End button on your phone, the person you're talking to goes bye-bye. While performing this technique, avoid pressing keys too quickly or you might accidentally cut your call short.

There's really only one step to this technique: Simply press the main menu key while on a call. This will let you access your phone's applications, but it won't end your conversation. I often use this technique to make a note of a phone number (using the Notes application), look up the name and number of a contact, or modify some contact information in the contact record of the person I'm talking to (if he's telling me his phone number has changed for example, I make the change in Contacts while he's on the line — how convenient is that?).

Technique 35: Making Conference Calls with Your Smartphone

Businesspeople already know the power of conference calling (where more than two parties are talking together at the same time). In a conference call, groups can come to agreement, make decisions, and plan for the future without having to hold a meeting where everyone is physically in the same place.

Even if you aren't a businessperson, you'll probably find a use for conference calling. School coaches, for example, can conference-call to map out team practice schedules. Teenagers can use conference-calling for arranging when and where to meet. Parents can conference-call the whole family together to discuss weekend plans.

The main reason conference calling has been exclusive to business has mostly to do with phone infrastructure and equipment. Business telephone setups allowed for speakerphones and conference calling bridges and the like that made conference calling possible.

But guess what? Your Series 60 smartphone supports conference calling. Here are the steps to make a three-way conference call:

1. **Place a call to your first party.**
2. **After you're connected to your first party, press Options on the left menu key.**
3. **Choose New Call (see Figure 5-24).**

Figure 5-24: Choose New Call on the Options menu to call your second party.

4. **Enter the number of your second party or Press Find on the left menu key and search in your Contacts directory.**

   During the connecting to your second party, your first call will be automatically placed on hold.

 Note   To switch between two calls, press Swap on the right menu key (see Figure 5-25).

Figure 5-25: Press Swap on the right menu key to switch between two active calls that are not yet conferenced.

5. **After you've made contact with your second party, select Options on the left menu key.**

6. **Choose Conference (see Figure 5-26).**

   All parties are now on the line.

Figure 5-26: Choose Conference to connect your two other parties and yourself and begin your three-person conference call.

 Ask your service provider's customer-service representatives how you will be charged for conference calls, or try one short conference call and see how it appears on your monthly bill. The usual scenario is that you're charged for each person you talk with as if it were a separate call. Therefore, you're on the line with two other people, which means you'll be billed as if you made two separate calls at the same time. If you have a lot of spare minutes in your plan, conference calls may make no change in your bill, but if you don't have many spare minutes, think before you start frequently using this technique.

 Not all service providers support conference calling even though your phone does. Double-check with your service provider's technical support if this tip doesn't work on your phone.

Series 60 phones support up to six parties on a conference call. Be advised, however, that your service provider might limit the maximum number to less than that. It's also important to point out that you can set up a separate call with one of the conference participants. Basically, you and the other party can have a private conversation while the other participants are muted and allowed to continue with the main call. Cool! It gets even better — you can also drop a participant from an active conference call.

When you start conference calling, you'll be hooked. It's the most efficient way for groups of people to make a plan together. Just watch those wireless minutes! Depending on your plan, conference calling could burn through minutes quickly.

Customizing Your Phone

Series 60–based phones offer a variety of ways to customize your phone to best suit your lifestyle and needs. Not only can you completely transform the look and feel of the screens using User Interface (UI) themes, but you can also move folders and applications around on your screens to put the items you use most often all within a few key clicks.

In this chapter, I show you how to change the format of your screen using new UI themes and different views. You'll also find out how to move folders and applications using the Move option. Finally, I describe the use of two cool downloads — Smartlauncher and Full Screen Caller — that you use to dramatically alter the look and feel of your phone.

Technique 36: Formatting Your Screens (List View versus Grid View)

By default, Series 60 phones come set to Grid view. This view makes your phone's main menu look a little like the Desktop in Windows. You'll see a set of icons representing applications (see Figure 6-1). You can select and click on these icons to launch applications on your phone. As always, navigation is handled with the convenient five-way (up, down, left, right, center-press) controller.

The Grid view works well if you simply like the look of the icon pictures or if you only have a few applications that you use frequently. With just a few popular applications, you won't need to scroll through several screens (after you perform Technique 37).

Figure 6-1: The Grid view of Series 60 phones makes the main menu look a little like the Desktop in Windows.

But if you couldn't care less about the pictures or you have a larger number of applications that you frequently use, consider the List view. The List view works just like it sounds. It turns the main menu into one long list of applications (see Figure 6-2). The list's order comes from the ordering of the applications in Grid view — so it isn't alphabetized.

Figure 6-2: List view turns your main menu into one long list of applications.

The List view makes for fast scrolling up and down to particular applications. You don't need to worry about left or right joystick clicks that you need to use in Grid view.

Tip For fast scrolling in List view, press and hold the joystick up or down. The scrolling keeps moving upward or downward automatically, without any additional key clicks. When you see the application you want to launch or the folder you want to open, simply take your finger off the joystick and the scrolling will stop.

Tip The 1 through 9 application shortcut keys described in Technique 25 still work while in List view.

Ultimately the decision to use either List view or Grid view is a matter of personal taste. Perhaps best of all, switching between the two views is very easy, so you can try both for a while and choose the one you want to keep very easily.

To switch from Grid view to List view:

1. **On the main menu, select Options.**
2. **Choose List View (see Figure 6-3).**

   The display is updated automatically.

Figure 6-3: Choose List View from the Options menu to switch the look of your main menu.

 Note   The main menu is not the only menu that you can alter with the List and Grid views on Series 60 phones. In Series 60, the main menu is just like any other menu, so in fact, you can change views in any menu — like Tools, Settings, and so on. The only trick: You have to perform the steps in every menu you want to adjust. Changing views on the main menu only affects the main menu. All your other menus remain as they are unless you perform the steps described in this technique while viewing the menu.

Do you love the change? It's okay if you don't. Switching back is easy. To switch from List view to Grid view:

1. **On the main menu, select Options.**
2. **Choose Grid View (see Figure 6-4).**

   The display is updated automatically.

Figure 6-4: Choose Grid View from the Options menu to switch back to the original menu format.

In techniques later in this chapter, I show you other ways to transform the look and feel of your main screen.

Technique 37: Moving Applications and Folders in Your Main Menu

This is one of those techniques that causes people to say, "I didn't know I could do that." It's one of the features of Series 60 that's glossed over — if it's mentioned at all — in smartphone user manuals. Another reason for the surprise might be that on older cellphones you couldn't change the order of the menus, so people probably don't expect the feature on their new phones.

Like the views described in the preceding technique, you can use this technique to alter the look of your main menu and any other menu on your phone. The smartest use of this technique is to move the applications and folders you use most frequently to the locations nearest the top of the main menu. You can also use this technique to organize the applications and folders of your phone to better suit your lifestyle or work style.

When you organize and move the applications and files on your phone, you can simply use the folders that come standard on your phone — like Tools, Games, and Extras. (Due to the fact that Series 60 phone makers can customize the Series 60 User Interface, some Series 60–based phones have different names for these standard folders.) Or, you can create your own new folders. Here's how to create a new folder on your phone:

1. **On the main menu, select Options.**

2. **Choose New Folder (see Figure 6-5).**

 Tip   Remember on most phones the 0 (zero) key is the space key in text-entry mode. On the Siemens SX1, the 1 key is the space key. Make your folder names even more descriptive by entering more than one word as the folder name.

Figure 6-5: Choose New Folder on the Options menu to create a new folder

Tip

Need to enter a number in a folder name? You could press a key several times until the number appears. For example, to get a 1, you need to press the 1 key five times. But there is a faster way: Just press and hold the 1 key (or any other key 0 through 9), and the number will be entered in the text-entry field that is currently active.

Note

There is a 35-character limit on folder names. However, you'll notice other more-important restrictions as you create and name new folders.

In Grid view, mentioned in the preceding technique, the display is limited to seven characters per icon (see Figure 6-6). Any folder or application name that is longer is truncated and displayed with a trailing ellipsis (. . .) to show that the name is actually longer.

In List view, the display is limited to 14 characters per application or folder (see Figure 6-7). Longer names are truncated and appear with trailing ellipses.

Figure 6-6: In Grid view, names are limited to seven characters.

Figure 6-7: In List view, names are limited to 14 characters.

3. **Enter the name of your new folder (see Figure 6-8).**

   Make it as descriptive as you can. Keep in mind that there is a 35-character limit on folder names.

Figure 6-8: Enter your new folder's name.

4. **Press OK.**

   The main menu will now be updated with your new folder.

After you've created a new folder, you can move applications and other folders into it. The following steps show how to move an application into a folder. In this example, I move the screen-capture application ScreenTaker (described in Technique 83) into the new Dean's Magic folder:

1. **On the main menu, scroll to the application you want to move (see Figure 6-9).**

2. **Selection Options on the left menu key.**

3. **Choose Move to Folder (see Figure 6-10).**

Figure 6-9: Scroll to the application you want to move.

Figure 6-10: Choose Move to Folder on the Options menu.

You'll now see the Move To menu.

4. **Scroll down to the folder into which you want to move your application (see Figure 6-11).**

Figure 6-11: Scroll to the new destination folder on the Move To menu.

5. **Press OK on the left menu key**

6. **Verify that the application has moved from the old location to the new destination (see Figure 6-12).**

Figure 6-12: The application has been moved to the new folder.

Technique 38: Downloading and Using User Interface Themes

User Interface (UI) Themes are a fast and easy way to radically change the look and feel of the screens on your phone. UI Themes are available for download from the Web, and they can transform the colors and style of your menus as well as the pictures used as icons for your applications. You can also download the free Series 60 Theme Studio from the Forum Nokia Web site (www.forum.nokia.com) to create your own new, exciting, and personalized themes. Themes are called themes because the menu and icon designs generally follow one concept, like a band, a sport, a season, a particular holiday, and so on.

The official Series 60 theme functionality is only available on the most recent Series 60–based phones, namely those running Series 60 2nd Edition and later. How do you know if your phone has themes? At the time of this book's publication, the following available Series 60–based phones included support for UI Themes: Nokia 6260, Nokia 6600, Nokia 6620, Nokia 6630, Nokia 6670, Nokia 7610, Panasonic X700, and Samsung SGH-D710.

Perhaps the easiest way to tell if your phone has Series 60 themes is to look for a Themes icon on the main menu. Phones without support for UI Themes naturally do not have a Themes icon.

If you find your phone doesn't support UI Themes, you're not completely out of luck. You can download simple themes from companies like Trigenix. These themes aren't as comprehensive (meaning they don't impact as many of the menus, backgrounds, and icons) as the official Series 60 UI Themes, but they do make a nice alternative.

Note You might recognize the idea but not the name *themes.* That's okay. Some other applications use the term *skins* rather than *themes* when describing the mechanism through which people can modify the look and feel of software. In Series 60 Web sites, though, you will find the term *themes.* Generally, Series 60 UI Themes make more comprehensive changes to your phone than "skins."

Series 60 2nd Edition phones come with a handful of themes already installed. Service providers sometimes preinstall their own in addition to the default set that comes with the Series 60 User Interface.

You can download new Series 60 themes from a variety of places, including Mango Themes (`www.mangothemes.com`). Here are the steps to download a new theme.

After you have themes on the phone, you can preview and apply them to your phone. Here's how:

1. **On the main menu, select Themes (see Figure 6-13).**

Figure 6-13: Select Themes on the main menu.

Now you'll see the list of themes residing in phone memory. The one with the checkmark in front is the theme that is currently active (see Figure 6-14).

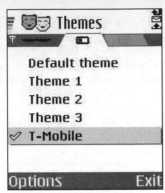

Figure 6-14: The theme with the checkmark is the currently active theme.

2. **Scroll to a nonactive theme to highlight it (see Figure 6-15).**

Note

To access themes stored on your phone's memory card, press the joystick to the right while viewing the phone memory themes. You can also jump to the special folder for theme downloads by choosing Options ⇨ Theme Downloads from the left menu key.

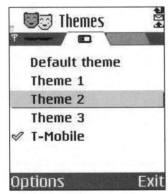

Figure 6-15: Scroll to the new theme you want to try.

3. **Press Options on the left menu key.**
4. **Choose Preview (see Figure 6-16).**

Figure 6-16: Choose Preview.

Now you'll see a preview of the new theme (see Figure 6-17). The title bar at the top of the screen shows the themes name. Notice the colors and the background picture. If the theme contains new icon pictures for applications and folders, you'll see them here as well. If you like the look of the theme, you'll want to apply it here.

Figure 6-17: Look at the preview to see if you like the changes in color and background.

5. **If you like the look of the theme, press Apply on the left menu key.**

   If the theme contains an update to the screensaver, you'll see a confirmation message asking if you want to replace your old screensaver with the new one (see Figure 6-18).

Figure 6-18: Choose whether you want to change the screensaver.

6. **Press Yes or No on the menu keys.**

Technique 39: Changing the Look and Feel of Your Phone's Menus Using the SmartLauncher Download

In the preceding technique, you discovered that only newer Series 60–based phones (Series 60 2nd Edition or higher) feature menu-transforming UI Themes. Did that news upset you? Are you ready to toss your older Series 60–based phone into the trash to make way for a new one?

Wait! As Douglas Adams says in *The Hitchhiker's Guide to the Galaxy,* "Don't panic!" You can still dramatically alter the look and feel of your older Series 60 phone's screens — and even change them frequently with other skins. The key is downloadable shareware called SmartLauncher.

 Note   SmartLauncher works on all Series 60–based phones.

SmartLauncher replaces your main menu's background and icons with a new skin (see Figure 6-19). You can choose from a set of default skins that come with the application, download additional skins created by other users, or create your own skin with freeware download Skin Editor (also from the creators of SmartLauncher). SmartlLauncher's change in "look and feel" may not cover as many menus as the official Series 60 UI Themes, but the transformations are still dramatic and cool.

Figure 6-19: SmartLauncher replaces your Series 60 main menu and icons.

But SmartLauncher doesn't just make over the looks — SmartLauncher also changes the Options menus (see Figure 6-20). Also note that the right menu key becomes Hide.

Figure 6-20: SmartLauncher also changes the Options menu of the left menu key.

But wait, there's more. As shareware developers often do, SmartLauncher also takes something good, namely the application shortcut menu (see Technique 27), and makes it better.

Press and hold the main menu key, and you'll see a world of difference between the original application shortcut menu (see Figure 6-21) and SmartLauncher's version (see Figure 6-22). In SmartLauncher's application shortcut menu, you'll see application memory usage for each active application and the free space that remains in your phone memory and your MMC (Multimedia Memory Card) if you have one in the phone.

Shutting off SmartLauncher

Nothing is as frustrating as installing some new application that dramatically changes your phone — and not knowing how to shut if off. Here's the way to turn off the SmartLauncher application, just in case you decide you can do without it:

1. On the main menu, press Options on the left menu key.

2. Choose Settings (see the figure).

Choose Settings on the
Options menu.

3. On the Settings screen, highlight Launcher.

4. Center-press the joystick to change the Launcher setting to Off.

You can also press Exit to close the SmartLauncher application. Setting Launcher to "Off" hides the application but keep it running in the background in case you reactivate it.

5. Press OK on the left menu key.

You'll still see the SmartLauncher screen as you return to the main menu. Do not be alarmed! I am a trained professional. Remember to keep your hands and feet inside the ride at all times as we move on to the next step.

6. Press the main menu key.

Now the main menu will turn back to the original Series 60 menu that you know and love.

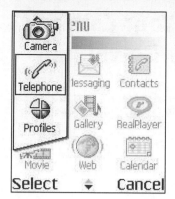

Figure 6-21: The original Series 60 application shortcut menu (see Technique 27 for instructions on how to use it).

Figure 6-22: SmartLauncher's application shortcut menu.

Okay, so how do you get SmartLauncher?

1. **Point your PC's Web browser to SymbianWare (**www.symbian ware.com**) and search for SmartLauncher.**

2. **Follow the site's instructions for downloading SmartLauncher to your PC.**

3. **Follow the instructions in Technique 72 for installing Symbian applications on your phone.**

Tip As with all the shareware mentioned in this book, try before you buy. SmartLauncher has a 15-day free trial. Try SmartLauncher on your phone before you spend the 10€/$13 (at the SymbianWare Web site) to purchase it.

Ultimately, though, SmartLauncher is a cool way to customize your phone's look and feel, especially for those phones that don't come with Series 60 themes.

Technique 40: Supersizing Your Caller ID Using the Full Screen Caller Download

The best shareware developers take something good and make it even better. That's exactly what the creators of Full Screen Caller did. They took a Series 60 feature that was already cool—images attached to contacts—and supersized it.

Instead of the standard thumbnail image used by the Series 60 interface, Full Screen Caller uses, well, a full screen to let you know who's calling (see Figure 6-23).

Figure 6-23: Full Screen Caller blows up your caller id.

The application uses your existing contact list. It allows you to assign pictures to contacts from your gallery of images or to use a new image captured directly by the camera. Here's an example of assigning a Gallery image to a contact in Full Screen Caller:

1. **On the main menu, select the Full Screen Caller application (see Figure 6-24).**

Figure 6-24: Select the Full Screen Caller application.

2. Select Pictures (see Figure 6-25).

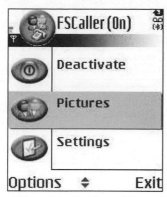

Figure 6-25: Select Pictures.

3. Scroll to the contact to which you want to assign a full-screen image (see Figure 6-26).

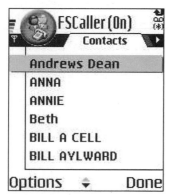

Figure 6-26: Scroll to a contact and select it.

4. Choose Picture.
5. Choose Select Picture (see Figure 6-27).

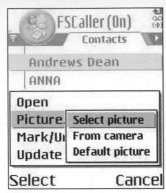

Figure 6-27: Choose
Select Picture.

6. **Navigate to the picture in your gallery.**

7. **Select the image (see Figure 6-28).**

Figure 6-28: Select the image
that you want to assign to
your contact.

Now, you'll see an image icon next to your contact's name.

Here's how to activate Full Screen Caller:

1. **Launch Full Screen Caller.**

2. **Select Activate (see Figure 6-29).**

Figure 6-29: Select Activate
to turn on Full Screen Caller.

By default, Full Screen Caller will display the large contact images both when
you receive a call from a contact and when you make a call to a contact. You
can adjust many settings regarding backlighting, colors, and image size under
Full Screen Caller's Settings menu.

 As always, try the Full Screen Caller using the free 15-day trial before you
buy.

Capturing, Editing, and Sending Video

◆ ◆ ◆ ◆

In This Chapter

Optimizing your
phone's video-recorder
settings

Troubleshooting
problems sending video
clips from your phone

Using video-playback
shortcuts on your phone

◆ ◆ ◆ ◆

Five years ago, it would have been hard to imagine shooting digital video with a mobile phone, let alone anything other than a clunky camcorder. Fortunately, cutting-edge engineers and software developers always strive to accomplish the unimaginable. Videophones are real, they're here, and they're wicked cool.

Guess what? You've automatically achieved cool status 'cause every Series 60 phone is a videophone. (The Nokia N-Gage and N-Gage QD can send, receive, and play video even though they have no camera to record video.) And if you're reading this book you must own or have access to a Series 60–based phone.

This chapter shows you the ins and outs of configuring and shooting video on your phone. You'll also learn tips and tricks for sending your video files in messages and e-mails. And you'll discover how to transfer video to your PC for viewing there.

Technique 41: Setting Up, Recording, and Sending Video

Seen the movie *Cellular?* It's a thriller about a kidnapped schoolteacher who makes one last desperate phone call. A Series 60–based phone, the Nokia 6600, is central to the plot. Even better, a video recording made by the phone saves the day — but I won't spoil the movie by telling you how, in case you haven't seen it.

Will you be ready to save the day with your phone's video recorder? Your video-recorder application should work fine right out of the box, but you'll get more out of your phone's video capability if you take a moment to understand and modify your video-recorder settings.

 Note Are your video-camcorder and digital-camera applications integrated together on your phone? It depends on which Series 60 phone you own. Newer phones, like Nokia's 7210 or Panasonic's X700, have a slightly newer version of the Series 60 platform, which integrates the digital camera and video camera together. This integrated camera and camcorder can be found on phones using Series 60 2nd Edition and later. In older Series 60 phones, the digital camera and video recorder application were separate items on the main menu. In the newer models (like Nokia 6260, 6620, 6630, 6670, and 7210, and Panasonic X700), when you activate the Cameras application, you'll see and have access to both the video recorder and the digital camera. To move from one to the other, just move the joystick to the left (see Figure 7-1) or to the right (see Figure 7-2).

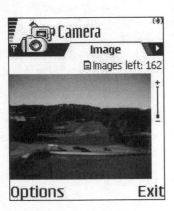

Figure 7-1: On newer Series 60 phones, like the Nokia 7610, the video recorder and the digital camera are integrated together. Go from one to the other by moving the joystick left and right. Here is the Camera.

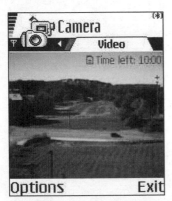

Figure 7-2: And here is the Video Recorder.

One important note: Video recorder settings changed between the initial Series 60–based phones (like the Nokia 7650, 3650, and N-Gage; Panasonic X700; Siemens SX1; and Sendo X) and later (2nd Edition) versions of Series 60 phones (Nokia 6260, 6620, 6630, 6670, 7610, and Samsung SGH-D710). So, you won't see on your phone every configuration setting listed here. That's okay. I list them all here, so that whatever phone you have, you'll learn what you need to know.

Tip
If you decide not to save a video just after you've recorded it, press the C key. You'll see a pop-up confirmation message asking you to confirm the deletion of the video clip you just recorded. Press Yes on the left menu key to complete the deletion.

✦ **Length (see Figure 7-3):** See the nearby sidebar "Having trouble sending a video clip from your phone?" for tips on adjusting this setting when recording a video clip that you want to send to someone. Select Short, and your video recording will stop automatically when you've reached 10 seconds. Select Maximum, and you'll continue recording video until you press Stop on the menu key or you run out of space in memory.

Figure 7-3: The Length setting tells the video recorder whether to stop recording after ten seconds.

Because I'm always ready to save the day with my video recorder, I keep this setting on Maximum so that the recording doesn't stop if I'm in the middle of recording something exciting. But if you send most of the video clips out from your phone, then you'll probably want to keep this setting on Short.

Note
What happens if you receive a call while you're recording a video? No worries. Your video recorder automatically stops recording and saves the video file you've recorded thus far under your default video filename. And, thanks to the power of multitasking, after you choose whether to accept or decline the incoming call, you're immediately returned to the now saved clip in the video-recorder application. This convenient trick works with all applications, not just the video recorder. Try that on a non–Series 60–based smartphone!

✦ **Video Resolution (see Figure 7-4):** As mentioned earlier, the lower setting (128x96) will make your video clips slightly smaller in size than the higher resolution. Also, the higher resolution (176x144) reduces the frame rate, making the final video appear slightly more jerky than a video created with the smaller resolution. So, this setting is a trade-off. Generally, I keep my video recorder set at the higher resolution because it looks better to my eye when I play it back.

Figure 7-4: The Video Resolution setting lets you set either low or high resolution.

✦ **Default Video Name (see Figure 7-5):** This is the name used as the prefix for any video clip you record. If there is another file with this name stored on your phone, the application appends a number (like 001, 002, 003), until it finds a filename that is unique. The default for this setting is "video." On my phone, I leave that as the default video name because I can't think of a more descriptive name, but you can change yours to anything you want.

Figure 7-5: The Default Video Name is the prefix used for every video clip you record.

✦ **Memory in Use (see Figure 7-6):** Here the choice is either Phone Memory or Memory Card. Because your phone's memory is shared by other applications on your phone—and there is no sure way to tell when you'll run out of phone memory—I keep this set to Memory Card.

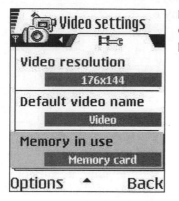

Figure 7-6: The Memory in Use setting determines if your video clips are stored in phone memory or on your memory card.

Tip

Press the Send button right after you record a video, and you'll see a pop-up menu asking if you want to send your video clip (see Figure 7-7).

Figure 7-7: The Send button is a shortcut key to send your video clip to another device.

There are two other video-recorder settings not under the Settings menu item inside the Video Recorder application:

✦ **Activate Night Mode (see Figure 7-8):** Activating this setting adjusts your video recorder for lower light conditions. It doesn't necessarily have to be night—it could just be inside your home or your office.

Figure 7-8: This video was shot using Normal Mode.

✦ **Mute (see Figure 7-9):** This setting turns off the audio portion of your video recording, making it a "silent film." It also makes for a slightly smaller video clip file in the end, as mentioned earlier.

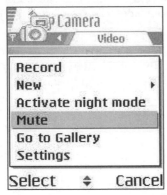

Figure 7-9: Use Mute to make a silent-film-style video.

Having trouble sending a video clip from your phone?

Have you ever taken a video clip with your phone and been excited about sending it to a friend or family member, only to have the send fail? The problem is probably the size of the video file. Phones themselves have limits on the size of files they can send out and service providers also have limits on the size of files they will route through their networks.

Here are some ways to reduce the size of your video files so that you'll have a better chance of sending them out to the world:

✦ On newer Series 60 phones (Nokia 6260, Nokia 6620, Nokia 6630, Nokia 6670, Nokia 7610, Samsung SGH-D710), you'll have a Length parameter in your Video settings. Set this to Short. This sets the maximum length of a video to roughly ten seconds.

✦ Set the Video Resolution to the smaller setting (128x96). On earlier Series 60 models (Nokia 3650, N-Gage, and 6600; Panasonic X700; Sendo X; and Siemens SX1), set the Image Size setting to L for the same effect. In a ten-second video clip, this only reduces the size by about 1KB, but on longer video clips the difference is greater. **Note:** Choosing the higher video resolution actually reduces the frame rate of your video recording, which means it will appear larger but slightly choppier than the lower resolution setting.

✦ On older Series 60 phones (Nokia 3650, N-Gage, and 6600; Panasonic X700; Sendo X; and Siemens SX1), find the Audio parameter in Video settings. Set this to Off. This turns off the audio portion of the video recording and reduces the overall size of your video file. On newer Series 60 phones (Nokia 6260, 6620, 6630, 6670, and 7610, and Samsung SGH-D710), use the Mute option under Settings to turn off the audio recording.

✦ Stop your recording earlier. Once, while traveling on business, I recorded a phone video of myself saying goodnight to my kids with my usual "sleep tight" bedtime wishes. The sending of this video clip failed because it was too big. So, I rerecorded the video with just a brief "Good night, kids. I love you." This video message arrived successfully in time for their bedtimes.

Technique 42: Using Video Playback Shortcuts for Fast Forward, Rewind, Mute, and More

Did you know you could scan through video clips forward or backward using keyboard shortcuts on your phone? You can also mute or raise the audio level while playing a video using shortcuts.

In this example, I walk you through all these shortcuts:

1. **On the main menu, choose Gallery.**

Note Your phone may have another name for the Gallery, where images and video are stored. If it's not obvious, check in your phone's user manual.

2. **Navigate to a folder that contains a video clip.**

Tip Want to find out more about your video clip file? After you find the file in the Gallery, choose Options ➪ View Details (see Figure 7-10) on the left menu key. The pop-up window will show you the size of the file, the date and time of last modification, the bit-rate and video format that was used, the resolution, and the name of the file (see Figure 7-11).

Figure 7-10: Choose Options ➪ View Details to find out everything you wanted to know but were afraid to ask about a video-clip file.

Figure 7-11: The gory details of your video file.

3. **Select the video-clip file.**

    This will launch the media player (Real Player on most Series 60 phones) to play your video.

4. While the video is playing, press the joystick up.

This will scan forward or fast-forward your video file. You'll notice that the video itself will freeze and the time readout (in seconds) in the upper-right-hand corner (see Figure 7-12) will increment rapidly.

Figure 7-12: The time readout in the upper-right-hand corner will fast-forward through your video as you press and hold the joystick up.

5. Now press the joystick down.

This scans backward or rewinds your video clip. You'll notice the time readout decrement for as long as you hold down the joystick.

6. Next press the joystick to the right.

This increases the volume of the audio soundtrack for your video. You'll notice the number of bars in the volume scale at the top of the screen rise as you hold the joystick to the right (see Figure 7-13).

Figure 7-13: Increase the volume during video playback by pressing and holding the joystick to the right.

7. Now press the joystick to the left.

This decreases the volume until you see the mute symbol (a speaker with a line through it) at the top of the screen (see Figure 7-14).

Figure 7-14: Mute your video playback by holding the joystick to the left.

Now you'll be able to fast-forward, rewind, and adjust the volume to a comfortable level while you play your favorite video files.

Playing phone video on your PC

Playing back video on your phone is fun, but let's face it, the screens are kinda small and the sound quality is good but not great. What can you do? Try playing the video on your PC. It's fun and once there, you can easily send the video to others via PC e-mail or post it on the Web.

For instructions on transferring files from your phone to your PC, read Technique 52.

Many standard media players for PCs have been or are now being updated to play the 3GP video format that is used on phones, including QuickTime 6.5 or higher (www.quicktime.com) for Windows or Mac and Windows Media Player 9 for Mac. So, after you've transferred your favorite video file to your PC, you can try opening it inside your favorite PC media player. If your media player doesn't support 3GP, it will say "format unsupported" when it tries to open your 3GP video file.

What to do? No worries! You've already installed your phone's PC software because you used it to transfer your file from your phone to your PC. All the PC suites for Series 60–based phones currently include a 3GP player. So, using Windows Explorer or another file system navigator, double-click on your phone video file to launch the 3GP player that was installed with your phone's PC suite software. See the figure for a look at Nokia's 3GP media player.

Nokia's 3GP video player for the PC.

Having More Fun with Your Smartphone

All work and no play makes Jack waste the poten-
tial of his smartphone. You've learned some great
productivity-enhancing, customization, and imaging
techniques in these chapters, but what if — as Cyndi
Lauper says — "girls just wanna have fun"?

With any powerful consumer-targeted technology, someone somewhere always thinks up a way to use it for fun
and entertainment. The same holds true for smartphones. In this chapter, you'll find out about companies
and software developers that have created e-book tools
for your phone, provided access to cameras all over the
world, and produced travel information that you can
use via your phone while you're on the go.

Technique 43: Receiving Pictures on Your Phone from Cameras around the World

You've heard about webcams, haven't you? Because of
the worldwide reach of the Internet and the natural ten-
dency of humans to share, we now can access cameras
all over the world by browsing the Web over our PCs.
Well, guess what? You can also access webcams using
your smartphone. Here's how:

1. **On your main menu, select Web (Services on
   pre–Series 60 2nd Edition–based phones
   from Nokia only, Internet on Siemens SX1,
   WAP on Sendo X for WAP sites, Web on
   Sendo X) to launch your phone's browser.**

Note Have you configured your phone to use the Web browser yet? If not, read Technique 2. It shows you how to request a special-configuration SMS message from your phone manufacturer or service provider to automatically configure your phone with the correct settings.

Tip Shortcut alert! On Nokia's Series 60–based phones only, in standby mode you can press and hold the 0 key to launch the browser.

2. **On the browser screen, press Options on the left menu key.**

3. **Choose Go To Address (on the Nokia 7650, 3600/3620/3650/3660, N-Gage, and N-Gage QD; the Siemens SX1; and the Sendo X only). On Series 60 Platform 2nd Edition and later phones (Nokia 6260, Nokia 6600, Nokia 6620, Nokia 6630, Nokia 6670, Nokia 7610, Panasonic X700, Samsung SGH-D710), choose Navigation Options, and click right (see Figure 8-1).**

Figure 8-1: Choose Navigation Options on your browser in newer Series 60–based phones.

4. **On the pop-up menu, choose Go To Web Address (see Figure 8-2).**

Figure 8-2: Choose Go To Web Address.

5. **On the address bar, enter** http://permodia.wilabs.com/wap **(see Figure 8-3).**

Figure 8-3: Enter the Web address for webcams.

6. **Press Go To on the left menu key.**

   You'll see a Welcome to Permodia WAP Camera List screen.

7. **Before going any further, save this page as a bookmark by pressing Options ➪ Save as Bookmark on the left menu key (see Figure 8-4).**

Figure 8-4: Save the page as a bookmark before you access the cameras.

8. **Next on the Welcome page (see Figure 8-5), select the text entry box by pressing down.**

Figure 8-5: Permodia's welcome page

9. **On the list of selections, choose Public Camera List (see Figure 8-6).**

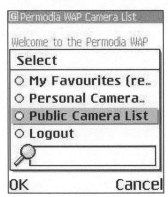

Figure 8-6: Choose Public Camera List.

10. **Press OK on the left menu key.**

You'll see a Select a Camera list box.

11. **Select the camera by pressing down on the joystick.**

    Now, you'll see a long list of webcams from all over the world (see Figure 8-7). Some of the names are a little inscrutable (Factoria), but most are obvious (Las Vegas, NorwayTraffic, Frankfurt, BigBen, and so on).

Figure 8-7: Permodia's list of world cameras

12. **Choose one by highlighting it and pressing down on your joystick.**

    The camera will snap a picture and display it in your browser (see Figure 8-8).

Figure 8-8: An instant picture from New York City's Times Square

Conversation starter? Ask the next person you flirt with where in the world he or she would like to be right now. Then instantly show him or her a picture from that place on your smartphone. I promise you'll appear worldly, high-tech-savyy, and, yes, cool!

Technique 44: Creating E-books and Reading Them on Your Cellphone

Okay, I'll admit it: I'm crazy for e-books. I was one of the, I guess, about three people on earth who actually purchased copies of e-books to read on my PDA. Unfortunately, it's a technology that hasn't caught on yet in the mainstream. BarnesandNoble.com, a major online bookseller, recently stopped selling e-books because, they said, there was just no revenue from it. For now, people still prefer reading books on paper.

So why this tip? Well, even if you prefer reading books printed on paper, I'm sure there are times when you're standing in line somewhere with nothing to read. By following the instructions in this technique, you'll never again be without something to read, as long as you bring along your smartphone. And, best of all, you'll be able to choose from thousands of free classic books to read on your phone. The only fee: a small charge for the two applications you'll need.

SymbianWare (www.symbianware.com) offers the two shareware applications you'll need for creating and reading e-books: ebook and ebookmaker. Point your PC Web browser to SymbianWare and follow the instructions for downloading these applications. Then use Technique 72 to install both of these Symbian applications on your smartphone.

While you're visiting the SymbianWare Web site, you should probably also download and install a sample e-book—the classic *Lady Chatterley's Lover* by D. H. Lawrence (see Figure 8-9).

Figure 8-9: *Lady Chatterley's Lover* as viewed through the ebook application on a smartphone

The applications are aptly named. Ebookmaker creates e-books out of text files right on your phone. Ebook lets you read the e-books (see Figure 8-10) that ebookmaker creates. Ebook is a well-designed reader application that lets you adjust font sizes and move page by page or line by line through a book. You can jump to specific chapters or set your own bookmarks to later start where you left off.

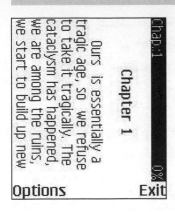

Figure 8-10: Catch up on your reading while stuck in line at the bank!

Tip For the most comfortable reading on a small screen, make the following adjustments on the Options ➪ Settings menus of ebook: Make the font size large, set the backlight on, and set the layout to landscape (rotates the text 90 degrees to give you a wider page).

Tip Ebook features automatic scrolling. Adjust the scrolling speed to match your reading speed, and then read page after page without even touching a key on your phone.

So, you're probably asking yourself, what about those free classic books he mentioned at the beginning of this technique? I haven't forgotten them. Point your PC Web browser to Project Guttenberg (www.guttenberg.net; see Figure 8-11). This massive archive stores text copies of thousands of classic fiction books. Unlike the recent events surrounding the illegal downloading of copyrighted music, no one will arrest you for downloading books from this archive. There is nothing illegal about your personal use of these classic works of fiction.

From the archive, select a book you want to read and download it to your PC following the instructions on the site. The book, you'll notice, arrives in text-file format — exactly the format the ebookmaker needs to create an e-book.

Transfer the book to your phone using the instructions in Technique 52. Then, launch ebookmaker and follow the instructions to make an e-book out of your text file.

Finally, launch the ebook application, and browse to your new e-book, following the instructions inside ebook.

Now begin reading!

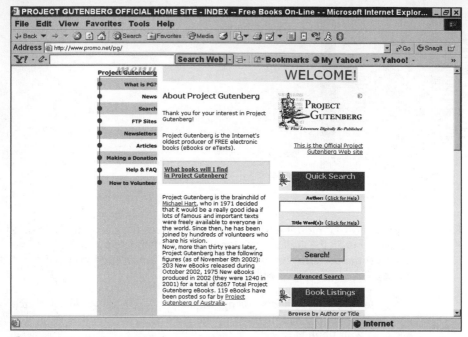

Figure 8-11: Project Guttenberg's archive features thousands of free classic books in text-file format.

Technique 45: Browsing Travel Information on Your Phone

Ever hear of the Lonely Planet travel guidebooks? Lonely Planet produces great travel information on both common and uncommon travel destinations. Best of all, the information is always up-to-date and relevant.

Did you know you have access to Lonely Planet's travel info right over your smartphone? Didn't think so. Here's how:

1. **On your main menu, select Web (Services on pre–Series 60 2nd Edition–based phones from Nokia only, Internet on Siemens SX1, WAP on Sendo X for WAP sites, Web on Sendo X) to launch your phone's browser.**

2. **On the browser screen, press Options on the left menu key.**

3. **Choose Go to Address (Nokia 7650, 3600/3620/3650/3660, N-Gage, N-Gage QD; Siemens SX1; and Sendo X only).**

4. On Series 60 2nd Edition and later phones (Nokia 6260, 6600, 6620, 6630, 6670, 7610; Panasonic X700; Samsung SGH-D710), choose Navigation Options, and click right (refer to Figure 8-1).

5. On the pop-up menu, choose Go to Web Address.

6. On the address bar, type http://mobile.lonelyplanet.com.

   Now, you'll see Lonely Planet's homepage (see Figure 8-12).

Figure 8-12: Lonely Planet's home page as seen through a smartphone Web browser

7. Highlight the link "Top places to eat, shop, and sleep."

8. Select the link.

   Next you'll see an A-to-Z list called Choose a City (see Figure 8-13).

Figure 8-13: Choose the letter of the alphabet that begins the name of your city.

9. Select the range that includes the city that interests you.

10. Select your city from the list of cities (see Figure 8-14).

Figure 8-14: Select your city.

At this point, you'll see the main page of your destination city (see Figure 8-15).

Figure 8-15: For each city, you'll find information on sites, hotels, and restaurants.

11. **Browse the overview or focus in on restaurants, hotels, or sites.**

    You'll find the information extremely valuable and very current.

Technique 46: Getting Video Highlights of Sports and Entertainment on Your Phone

Browsing the Web on your phone is impressive. But if you really want to blow your friends and family away, show them video highlights from a sporting event or a red-carpet runway of a film premiere on your phone.

In 2004, some lucky Series 60 phone users could order video highlights of the Summer Olympics and U.S. NBA season sent to their Series 60–based smartphones through some service providers. Look for more such sporting-event-based phone video highlights in 2005 and beyond.

1. **In the meantime, point your PC browser to PocketCinema (www.pocketcinema.com), shown in Figure 8-16.**

Figure 8-16: Browse PocketCinema for phone video highlights of sports and entertainment.

2. **Click the Downloads link.**

3. **Select PocketCinema Entertainment and Download Service from the list.**

4. **Follow the instructions to save the zip file to your PC hard drive.**

5. **Read Technique 73 and follow the instructions to install the Java JAR files for Portable Hollywood and Action Sports to your smartphone.**

6. **After you've completed the installations, on your phone's main menu, select Portable Hollywood.**

   You'll see the main screen of Portable Hollywood (see Figure 8-17).

Figure 8-17: Portable
Hollywood's main screen

7. **Browse the latest entertainment news by pressing your joystick to the right.**

   In addition to ring tones, images, and gossip, each weekly issue of Portable Hollywood offers a (currently) free video download of entertainment news.

8. **Press the 0 key on the video-download screen and follow the instructions to download the highlight video to your phone.**

9. **To have a weekly SMS of Portable Hollywood sent to your phone, press Subscribe on the right menu key and follow the instructions.**

 Tip   Make sure you double-check the cost of services like PocketCinema. Sometimes services start free and then begin to charge money over time.

Follow the same instructions to download the latest sports highlights and/or subscribe to Action Sports.

Technique 47: Playing MP3 Music on Your Phone

Apple's iPod may be a top-selling portable digital-music player, but think twice before you rush out and purchase one. You may want to explore your smartphone's music capabilities before you add another expensive device to your list of gadgets.

 Note   Not all Series 60–based phones support MP3 and AAC format music files. This technique really only applies to Series 60 2nd Edition–based phones like the Nokia 6260, 6620, 6630, 6670, and 7610, and the Samsung SGH-D710. Read your phone's user manual to find out whether your phone plays MP3 and AAC music.

Your smartphone features an integrated music player (see Figure 8-18) that supports MP3 and AAC digital-music formats — the two most popular digital-audio formats around.

Figure 8-18: The Series 60 music player supports MP3 and AAC digital-music formats.

You can even create play lists (like iTunes and iPod) on your PC or your smart-phone and share them back and forth between the two.

Follow these steps to create a play list on your PC and download it to your smart phone:

Note   Read Technique 49 for information on finding the PC suite software that matches your particular smartphone. Each phone manufacturer has its own version of PC suite software. In this example, I use Nokia's PC software, but most Series 60–based PC suite software works in a similar way (even from different manufacturers).

1. **Download your music tracks from the Web or convert them into digital format from CD.**

Note   Digital Rights Management (DRM) technology makes it impossible for you to play tunes you've downloaded from some music services — like iTunes or MusicMatch and play them on your phone. The licensing you've agreed to when you signed up with these services restricts the number and types of devices on which you can play the tunes.

Tip   Use shareware or integrated CD conversion (also known as *ripping*) soft-ware to convert CD audio into MP3 or AAC formats.

2. **Find the tracks on your PC using Windows Explorer. Search for all tracks of a certain format by entering *.mp3 or *.aac as the search string.**

3. **Launch your PC phone software (see Figure 8-19).**

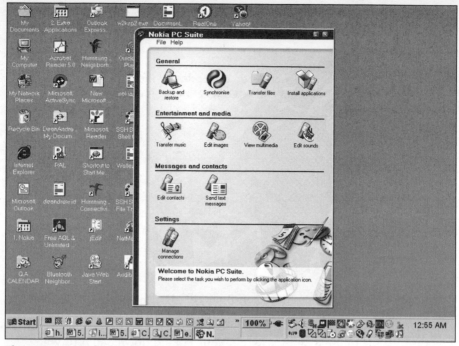

Figure 8-19: Nokia's PC Suite software for the Nokia 7610

4. **Launch the Transfer Music application (see Figure 8-20).**

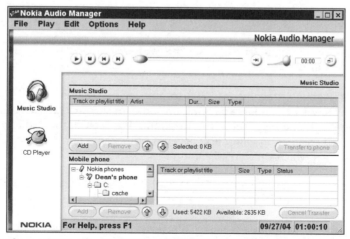

Figure 8-20: The Audio Manager for the Nokia 7610

5. **Create a new play list on your PC.**

   For the Nokia 7610 PC Suite Audio Manager, choose Edit ➪ Create New Playlist.

6. **Enter a name for your play list (see Figure 8-21).**

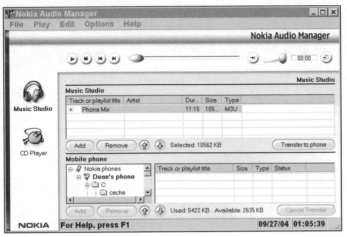

Figure 8-21: Create a new play list.

7. **Add tracks to your new play list.**

   In this example, press the Add button and browse to music tracks on your hard drive (see Figure 8-22).

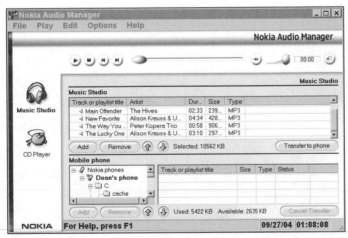

Figure 8-22: Browse to music tracks on your hard drive.

8. **Transfer the play list and your tunes to your phone.**

In this example, highlight the play list and press the Transfer to Phone button.

Tip

Music tracks generally run several kilobytes in size. If you have a Multimedia Memory Card (MMC), store your tunes there rather than in phone memory.

Now, your new play list and tunes are available on your phone. Who needs a dedicated digital-music player?

Connecting Your Phone to Your PC, Laptop, PDA, and Other Devices

The ability to connect your phone to a PC and other devices (for example, phones, PDAs, and so on) is the key difference between your smartphone and older cellphones. The ability to connect opens up a world of opportunity. *Carpe diem!*

With a PC or device connection you can synchronize contacts, e-mail, and other files; transfer images, video, and music between your phone and other devices; download and install applications from the Web; and much more.

If you've never connected your smartphone to a computer, sit down right now and try it following the steps in this chapter. It's not as hard as you think, and the benefits of doing it are tremendous. In this chapter you'll also learn how to find the right PC software for your phone, how to transfer files between devices, how to exchange virtual business cards with others, and how to use your phone as a modem for your PC or PDA.

Technique 48: Exchanging Business Cards (Contacts) with Other Phones

Traditional paper business cards are now officially obsolete! Well, maybe this statement is more of a predication than a proclamation, but by learning this technique, you'll see the amazing benefits of exchanging virtual business cards rather than paper ones.

What is a virtual business card? It's really a contact record in your phone's Contact database. It contains all the fundamental contact information that a paper business card contains — name, title, address, phone, fax number, and mobile number, e-mail address, and so on. The only item missing is the company logo. When you go to send contact information to someone, you'll notice your phone actually refers to the information as a "business card."

The first thing to know: Exchanging virtual business cards is fast and easy to do. You can exchange contact information from your phone to another phone in seconds using either Bluetooth or an infrared connection, or by sending an SMS or MMS message.

Plus, it's a great way to cut down on data entry. The main problem with a paper business card is that you have to enter all the data on the card into your computer or your phone, unless you like the fat-wallet look you get by carrying around other people's business cards. By exchanging electronic cards, there is no data entry. You can save the business card on your phone, and then synchronize your phone with your PC. And — *voilà!* — your contact information is up-to-date, and you never had to sit down and type in someone's name, address, phone number, Web site, cell and fax numbers, or anything else.

Follow these steps to exchange a virtual business card with another phone:

1. **Make sure that the device you plan to connect to supports the connection method (Bluetooth, infrared, and so on) you intend to use, and ensure that that capability is activated on both devices.**

2. **On the main menu, select Contacts.**

Tip

> In Standby mode, a center press of the joystick opens the Contacts application on all Series 60–based phones.

3. **Choose a particular contact from your list of contacts.**

4. **In the contact record, select Options on the left menu key.**

5. **Choose Send (see Figure 9-1).**

6. **Choose your sending method from the list (see Figure 9-2).**

    You can choose text message, multimedia message, Bluetooth connection, or infrared connection.

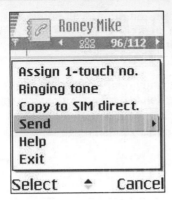

Figure 9-1: Choose Send to send a virtual business card of a contact record to another phone.

Figure 9-2: Choose the method to send your virtual business card.

 Note

If you already have an item in someone's contact record highlighted, the send procedure will ask if you want to send only the highlighted information or send all the contact's information (see Figure 9-3). In general, I try to send the whole contact record, but before you do, think about whether you have any information in the record — home phone number, perhaps — that the person might not want shared. Also, you may want to consider whether the recipient's device can support multiple fields within an entry.

Figure 9-3: If you have an item (like Mobile) highlighted within a contact record, you'll be asked if you want to send only the highlighted information or the whole contact record.

- **If you choose to send via Bluetooth,** the phone will immediately scan for devices within range (see Figure 9-4). Choose the recipient's phone from the list of Bluetooth devices to complete the sending of the business card.

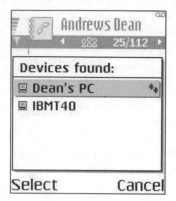

Figure 9-4: When you select Bluetooth, the phone will immediately scan for other Bluetooth devices within range.

- **If you choose to send via text message,** this throws you into the text message editor with the contact information already placed in the message body (see Figure 9-5). Choose Options ➾ Add a Recipient (or just center-press the joystick while in the address field to open the Contacts application) to select the recipient from your list of contacts, or you can enter the recipient's mobile phone number directly. Then, send the message as you normally would.

Figure 9-5: Choosing Text Message brings you into the message editor with the business-card information ready to send.

- **If you choose to send via multimedia message,** this action puts you into the multimedia message editor. You can add a text message like "Here's the number you wanted." The business card has been automatically added as an attachment. To view the business-card attachment, choose Options ➪ Objects on the left menu key (see Figure 9-6). Choose Options ➪ Add a Recipient (or just center-press the joystick while in the address field to open the Contacts application) to select the recipient from your list of contacts, or you can just enter the recipient's mobile phone number directly. Then, send the message as you normally would.

Figure 9-6: The business card will be added as an attachment to the multimedia message.

- **If you choose to send via infrared,** the phone will immediately scan for devices within range (refer to Figure 9-4). Make sure your recipient's phone has IR activated and that its IR port is facing your phone's IR port. Also, make sure the distance is only 1 or 2 inches apart for the best connection.

Now you know how to send a business card. Follow these steps to save a business card after you receive one:

1. **On the main menu, select Messaging.**

2. **Select Inbox from the list of folders.**

   The business card you received will show the sender's name (only if the sender is one of your contacts) or phone number (see Figure 9-7).

3. **Select the business card in your inbox.**

4. **Inside the business card (see Figure 9-8), press Options on the left menu key.**

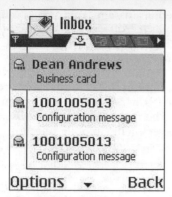

Figure 9-7: The business card will appear in your inbox.

Figure 9-8: An open business card

5. **Choose Save Business Card (see Figure 9-9) from the list of options.**

Figure 9-9: Choose Save Business Card to store the information in contacts.

The new contact information is now stored in your contacts directory.

Virtual-business-card etiquette

Here's the funny thing about giving someone a virtual business card: It's high-tech, but it's still personal. Two of the four possible methods to give someone a business card – a Bluetooth or an infrared (IR) connection – require you to be physically close to the person. Thus, there is some etiquette involved.

First, you have to agree on the method to use for exchanging business cards. For a Bluetooth connection, both phones need Bluetooth capability (all current Series 60 phones feature Bluetooth) and it must be active (see Technique 50 for more information on Bluetooth activation). For IR, both phones need IR ports (not all Series 60 phones feature IR ports – the Nokia 7610, N-Gage, and N-Gage QD do not, for example) and they both must be active (see Technique 51 for information on how to activate IR).

When I offer someone an electronic business card, I always let him choose the method we use. Some people might know how to activate their IR port, for example, but they might never have tried to use Bluetooth – so setting up a Bluetooth connection between phones could become a source for embarrassment. I just make sure I know all the methods of communication so I'm ready to use any of them.

Make sure you create a contact record with your own information – name, phone numbers, e-mail address, and so on – in the contacts application of your phone. This way, you'll be ready to trade virtual business cards with others in a meeting.

Another common use of this technique is forwarding business-card information to others. Someone in a meeting is looking for the phone number of someone else who isn't in the meeting. If I have the person's number stored in my phone, I pass the person a virtual business card containing the information he wants.

Technique 49: Finding and Setting Up Your Phone's PC Software

PC software empowers your phone. It's the main mechanism connecting your phone and PC. Through it, you can synchronize contacts, e-mail, and calendar appointments, transfer images from your phone to your PC, install applications and games on your phone, download music to the phone, and more. In truth, PC software makes your PC a portal out to the world for your phone.

Every Series 60 phone has PC-based software that goes with it. Some phone makers put the software on a CD-ROM right in the box with the phone. Others require you to download the software from the Web. Whichever is the case with your phone, find the software, download it, and install it. The effort will be well worth it.

Where to find your phone's PC software

Go to the appropriate Web site to find information on your phone's PC software:

+ **Nokia:** www.nokia.com

+ **Samsung:** www.samsung.com

+ **Panasonic:** www.panasonic.com

+ **Sendo:** www.sendo.com

+ **Siemens:** www.siemens-mobile.com

Sendo and Siemens offer something else special on their Web sites. With a special download, you can completely upgrade the firmware on your Sendo X and Siemens SX1. As of this book's writing, Sendo and Siemens were the only Series 60 phone makers offering this capability. If you own an X or SX1 take advantage of this free opportunity to keep your phone's software current and visit the Sendo X pages or Siemens SX1 pages on the manufacturer Web sites to download the firmware upgrade.

Your challenge is finding and installing the correct version of PC software for your phone. Nokia, for example, has several versions of PC software called PC Suites. Only one version works with a particular phone. Fortunately, Nokia supplies an online tool on its Web site that helps you match your phone model to a particular PC Suite.

Download the PC software to your PC's hard drive and follow the instructions to install it.

Tip If you've never used phone PC software before, look for tutorials, which many phones offer, and download them as well.

Finally, make sure you also retrieve the PC software's user manual.

Technique 50: Connecting to Your PC via Bluetooth

Bluetooth is truly an amazing technology. Using a radio frequency, it allows devices within 10 meters of each other to create an instant wireless network. Over this network, devices share data, but for smartphone users the data shared are messages, images, video, music, ring tones, calendar appointments, e-mails, applications, games, and more.

The smartphone community did not invent Bluetooth, but you can thank them for delivering it out to the world of consumer devices. Currently, every Series 60 phone on the market supports Bluetooth connectivity. Better still, you can use a Bluetooth connection between your phone and PC

Making your PC Bluetooth-ready

Does your PC speak Bluetooth? Many newer Apple iMacs come with integrated Bluetooth modules. Several PC notebook computers also feature Bluetooth. Generally, desktop PCs are lagging slightly behind, but some newer models do have integrated Bluetooth chipsets. Check your computer's user manual to find out whether it delivers Bluetooth network capability.

Even if your computer doesn't have Bluetooth, it's not the end of the world. You can purchase an inexpensive Bluetooth add-on that makes your PC Bluetooth-ready. Some snap into a PC card (also known as PCMCIA) slot; others plug into the PC's USB port. Examples of some available offerings are:

✦ Belkin's USB Adapter (www.belkin.com)

✦ D-Link Wireless USB Bluetooth Adapter (www.dlink.com)

✦ 3Com Wireless Bluetooth PC Card (www.3com.com)

Follow these steps to connect your PC and phone via Bluetooth:

 Note

Phone PC software, as mentioned in the last technique, varies from phone to phone. In this example, I use the Nokia's PC Suite Version 6, which supports Nokia's 7610 phone. You may not see exactly the same icons in your PC software, but all the same concepts apply and all PC software for Series 60 phones work in a similar way. Download and read the user guide that goes along with your particular software for specific details.

1. **On your PC, make sure you have Bluetooth enabled.**

   If you have integrated Bluetooth, access your Control Panel and enable Bluetooth. Or, if you're using an add-on Bluetooth module, make sure that it's plugged in and active.

2. **On your phone's main menu, select Connectivity (see Figure 9-10).**

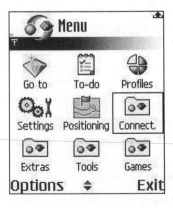

Figure 9-10: Select Connectivity on your phone's main menu.

Configuring your smartphone's Bluetooth settings

Access your phone's Bluetooth settings, on your PC, by selecting Connectivity ⇨ Bluetooth on your phone's main menu. Here's a breakdown of what the settings mean and advice on how to set them:

✦ **Bluetooth On/Off:** Turn this on to enable Bluetooth personal networking.

✦ **My Phone's Visibility:** This must be set to Shown to All in order for you to make a connection between your phone and PC using Bluetooth.

✦ **My Phone's Name:** Enter a descriptive name like "Dean's Phone." This is the name that will appear on the PC or other device during device "discovery."

3. **On the Connectivity menu, select Bluetooth (see Figure 9-11).**

Figure 9-11: Select Bluetooth.

4. **Confirm the following Bluetooth settings:**

   • Bluetooth: On

   • My Phone's Visibility: Shown to All

   • My Phone's Name: Some name that will be easy for you to identify when your PC is searching for Bluetooth devices within range during device discovery

5. **On your PC, launch your phone's PC software.**

6. **On the PC software's main menu, select the Connection Manager (Settings) (see Figure 9-12).**

7. **Select the appropriate Bluetooth connection.**

   In this example, I use the common Windows Bluetooth connection called WIDCOMM (see Figure 9-13).

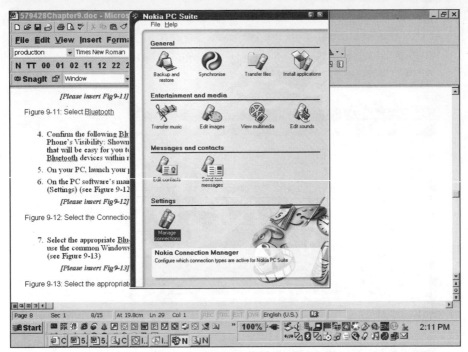

Figure 9-12: Select the Connection Manager of your PC software.

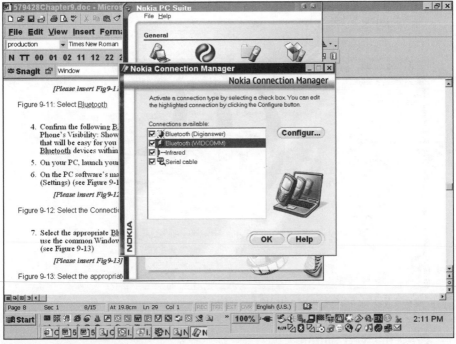

Figure 9-13: Select the appropriate Bluetooth connection for your PC.

8. **Press the Configuration button.**

9. **Press the Next button on the Welcome screen.**

Your PC will begin device discovery, searching for Bluetooth devices within range (see Figure 9-14).

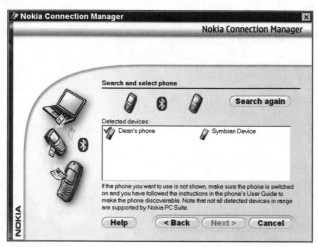

Figure 9-14: Your PC software searches for Bluetooth devices within range.

10. **Select your phone from the list of discovered Bluetooth devices (see Figure 9-15).**

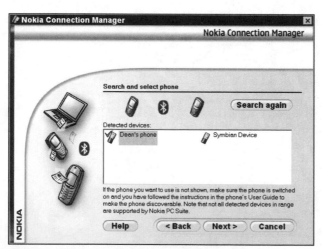

Figure 9-15: Select your phone from the list.

11. **Press the Next button.**

12. Enter a numerical passcode for this connection (see Figure 9-16).

This is a security measure so that others with Bluetooth can't make connection to your phone without your permission (this unwanted access is called *BlueSnarfing* and is covered in more detail in Technique 62). I suggest a sequence of at least four digits that is not obvious (at least not as obvious as 1234 or 1111).

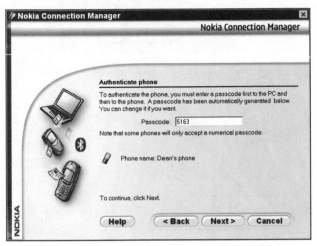

Figure 9-16: Enter a security passcode for your phone-to-PC connection.

13. Press the Next button.

14. Switch to your phone.

You'll see a Bluetooth pop-up entry box asking you to enter the proper passcode.

15. Enter the passcode (see Figure 9-17).

Figure 9-17: Enter the same passcode on the phone to make the Bluetooth connection.

16. **Switch back to your PC.**

17. **Press the Finish button on your Connection Manager screen (see Figure 9-18).**

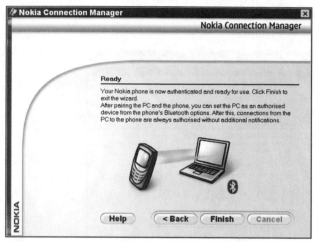

Figure 9-18: Press Finish.

18. **Press OK to close the Connection Manager (see Figure 9-19).**

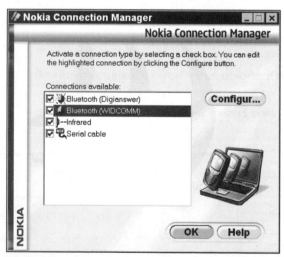

Figure 9-19: Press OK to close the Connection Manager.

Still with me? You successfully made a Bluetooth connection between your phone and PC. Congratulations!

Now, the following steps are not required, but it's a good idea to make a paired device association between your phone and PC. You'll probably be using this Bluetooth connection quite frequently to synchronize data and transfer images and files back and forth. After you've paired them, the devices will automatically make a connection between each other when they get within range of each other, provided Bluetooth is turned on in both devices.

1. **Switch back to the phone.**

2. **Inside the Bluetooth application, press the joystick to the right to move to Bluetooth's Paired Devices tab (see Figure 9-20).**

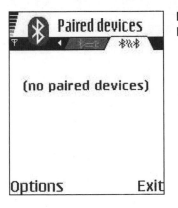

Figure 9-20: View the Bluetooth Paired Devices tab on your phone.

3. **Press Options on the left menu key.**

4. **Choose New Paired Device (see Figure 9-21).**

Figure 9-21: Make a paired device association between your phone and PC.

At this point, you'll see a list of nearby Bluetooth devices.

5. **If you don't see your PC, choose More Devices. Otherwise, pick your PC from the list of Bluetooth devices (see Figure 9-22).**

Figure 9-22: Pick your PC from the list of Bluetooth devices.

Now, your PC's name will be added to the list of paired devices (see Figure 9-23).

Figure 9-23: Your PC will appear on the list of Bluetooth paired devices.

Technique 51: Connecting to Your PC via Infrared

Overwhelmed by all the setup involved in a Bluetooth connection between your PC and phone? No worries! An easy alternative to Bluetooth for connections is infrared (IR). IR is a light-based technology and is the same technology used in your TV and DVD player remote controls. However, unlike your remote control, the IR port on your phone can be used for two-way communication. So, you can use IR to send and receive files, images, video, and music or to synchronize e-mail, contacts, and calendar to a PC.

 Tip   Most laptop or notebook computers feature integrated IR ports right from the factory. Most desktop PCs on the other hand do not. As with Bluetooth, you can add an IR port to your PC.

With IR, there's less setup than with Bluetooth, but it's less convenient because you need to keep your phone stationary next to your PC and also make sure that the IR ports are facing one another.

 Note   When I say an IR connection can be less convenient than Bluetooth, I'm not kidding. One Series 60–based phone that I own has an IR port on the left-hand side of the phone. My notebook computer also sports its IR port on the left-hand side of the case. This means when I place my phone next to my PC for an IR connection that my phone is upside down. What to do? Try to time your IR connections so that you won't need to move or use your phone (because it will need to remain stationary) for anything other than the data/information transfer. Also, if needed you can interrupt a IR connection — to answer a call, for example — but you will likely have to manually restart the IR connection after the interruption.

Again, read the instructions in your own phone's PC software regarding setup and use. Generally, though, the steps for making an IR connection between your phone and your PC are the following:

 Note   In this example, I'm using a Nokia 6600 phone and the Nokia PC Suite for the 6600 on the PC.

1. On the PC, launch your phone's Connection Manager.

2. Choose the Infrared (IR) port as the method of communication (see Figure 9-24).

3. On your phone's main menu, select Connectivity.

4. Select Infrared.

   You'll see a confirmation message that Infrared is activated. Shazaam! Your IR connection been made.

5. On your PC software check the status bar (generally in the lower-right-hand corner) to determine whether the PC software has successfully made the connection to the phone.

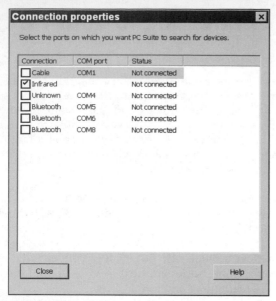

Figure 9-24: Choose an IR connection.

Technique 52: Using PC Software to Transfer Images, Synchronize Data, and More

Now that you've made a connection between your phone and PC, you've harnessed the tremendous power of a Series 60–based smartphone. You can now transfer images and synchronize e-mail, calendar, and contact info. You can also use your PC software to create SMS and MMS messages using your PC keyboard and image files. You can create digital music play lists and download them to your phone, view multimedia, and on and on and on. Also, you can use the Series 60 Theme Studio (available at Forum Nokia, www.forum.nokia.com) to create theme applications for your Series 60 on your PC and easily send the new file over to your phone. (*Note:* UI Themes are not supported in earlier Series 60–based phones like Sendo X and Siemens SX1, Nokia 7650, Nokia 3600/3620/3650/3660, N-Gage, and N-Gage QD.)

Follow these steps to transfer an image from your phone to your PC:

1. **If you haven't already done so, use Techniques 49 and 50 or 51 to install your phone's PC software and establish a connection between your phone and PC.**

Note   The look and feel of each phone's PC software varies slightly. In this example, I use Nokia PC Suite Version 6, which is compatible with the Nokia 7610 phone. Your phone's software will operate in a similar way. Please read the documentation of your phone's software for details.

2. **On your PC, launch your phone's PC software (see Figure 9-25).**

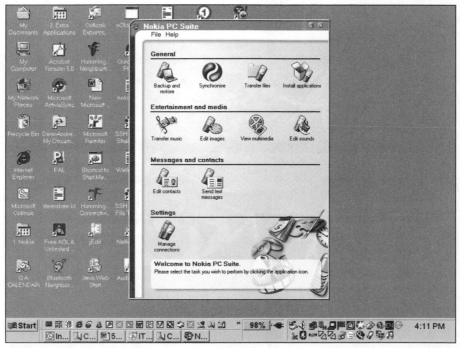

Figure 9-25: Launch your phone's PC software.

3. **Select Transfer Files.**

4. **On the file explorer window, select your phone (see Figure 9-26).**

   Now you'll see Phone Memory and Memory Card (if you have a memory card currently installed on your phone), as shown in Figure 9-27.

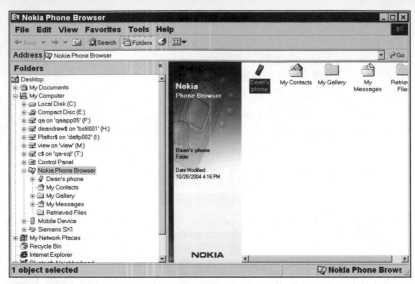

Figure 9-26: Select your phone in order to view the images and files inside it.

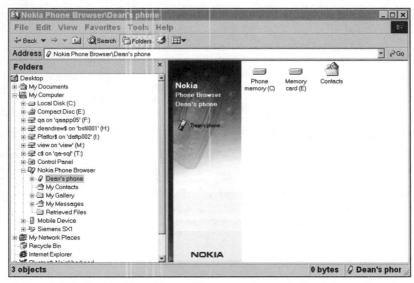

Figure 9-27: Select either Phone Memory or Memory Card depending on where your image resides.

5. **Select the one that contains the image you want to transfer from your phone to your PC.**

 Tip

On Nokia phones, the next two folders will be Cache and Nokia. Select Nokia to access images and other files.

6. **Select the Images folder (see Figure 9-28).**

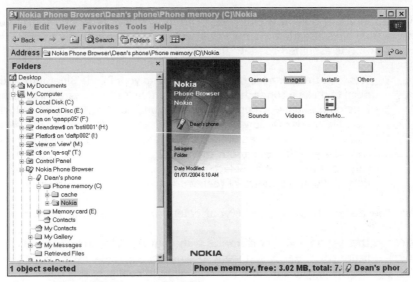

Figure 9-28: Select the Images folder.

7. **Select the image(s) you want to transfer to your PC (see Figure 9-29).**

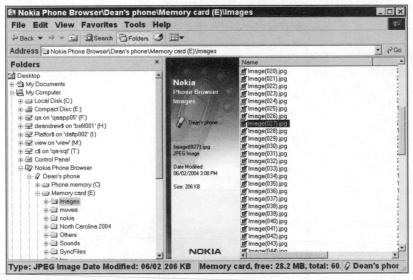

Figure 9-29: Select your images.

8. **Right-click with your mouse and choose Copy.**

9. **Using the file explorer, navigate to the folder on your hard drive.**

10. **Right-click with your mouse and choose Paste.**

Congratulations! You've now transferred an image from your phone to your PC. If you double-click the file on your PC, it will open automatically in whatever PC application is assigned to view JPEG image files.

Note To simply transfer a file from your phone to your PC via Bluetooth, you don't need to make a connection using your phone's PC software. First, make sure Bluetooth is active on both your phone and your PC. Then, while highlighting a file choose Options ⇨ Send ⇨ Via Bluetooth. Your phone will search for Bluetooth devices in the area. Select your PC. Your PC's Bluetooth software will alert you that Bluetooth contact has been made and it will ask you if you want to accept it. Select Yes. The files you sent will be stored in the default Bluetooth directory (defined by your Bluetooth software).

Follow these steps to synchronize your calendar between your PC and phone:

Note The first time you synchronize with your PC software, you'll need to configure what exactly you want to synchronize. These instructions will show this initial configuration. You only need to configure this once.

1. **On your PC, launch your PC software.**

2. **Select Synchronize (see Figure 9-30).**

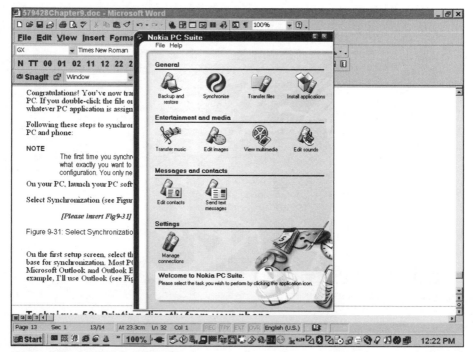

Figure 9-30: Select Synchronize.

3. **On the first setup screen, select the PC application that you want to use as the base for synchronization.**

   Most PC software lets you choose from among Microsoft Outlook and Outlook Express, or Lotus Notes and Organizer. In this example, I use Outlook (see Figure 9-31).

4. **Click the Next button.**

5. **Check the boxes of the elements you want to synchronize.**

   You can choose from Calendar, Contacts, To-do lists, Notes, and Email. For this example, I'm choosing Calendar (see Figure 9-32).

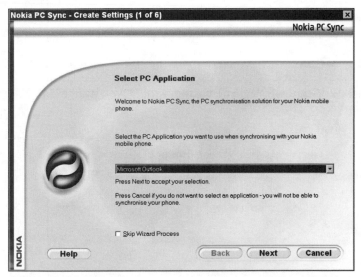

Figure 9-31: Select the PC application you want to use for synchronization.

6. **Click the Next button.**

7. **Using the Browse button, navigate to your Calendar folder (see Figure 9-33).**

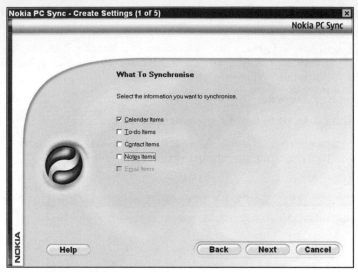

Figure 9-32: Select which elements from your PC you want to synchronize.

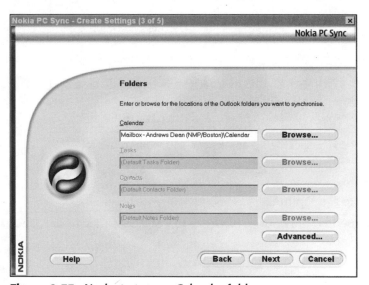

Figure 9-33: Navigate to your Calendar folder.

8. **Click the Next button.**

9. **Select the range for synchronize — how far into the past and future should the application synchronize your calendar appointments and reminders.**

I, typically, use the defaults — one week into the past, one month into the future (see Figure 9-34).

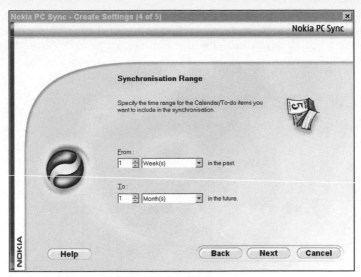

Figure 9-34: Choose the range of synchronization.

10. **Click the Next button.**

11. **You made it! Check the box asking whether you want to synchronize immediately, and then click the Finish button (see Figure 9-35).**

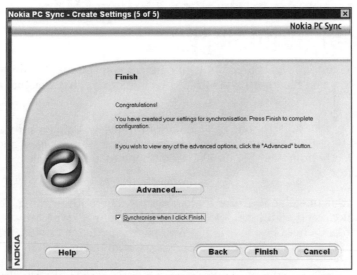

Figure 9-35: Woohoo! You've completed the synchronization configuration.

Now, anytime you want to synchronize (I suggest at least once every day), select the Synchronize application of your PC software and click the Synchronize Now button (see Figure 9-36). *Note:* The title of this button varies from PC software to PC software.

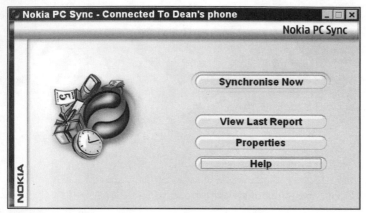

Figure 9-36: Synchronize whenever the mood strikes by clicking Synchronize Now.

Technique 53: Printing Directly from Your Phone to a Printer Using HP's Bluetooth Technology

Earlier in this chapter, I showed you how to connect your phone to another phone and a PC using Bluetooth and infrared technologies. That was just a warm-up!

Believe it or not, your smartphone can also *print* wirelessly. Using a software download and Bluetooth technology you can print pictures, messages, and documents directly from your Series 60–based phone to a compatible Bluetooth printer.

Hewlett-Packard is one manufacturer that allows for Bluetooth printing. For the software download and additional information, point your PC Web browser to Hewlett-Packard's Camera Phone Printing Web page (http://h41145.www4. hp.com/uk/en/phoneprinting/default.htm), shown in Figure 9-37.

Note   Not all printers offer Bluetooth communication capability. Check with your printer manufacturer regarding the capabilities on your printer. The HP DeskJet 450wbt is one example of a Bluetooth-capable printer.

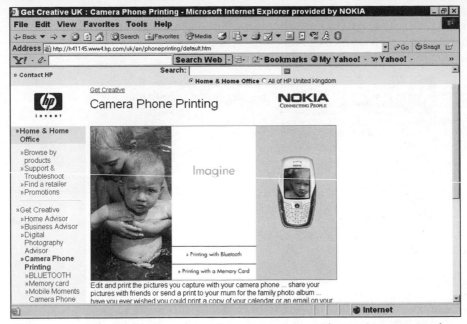

Figure 9-37: Print directly from your phone to a printer without wires. How cool is that?

 Tip

You can also find Series 60 Bluetooth printing software on Nokia's Web site (www.nokia.com). Navigate to the page for the Nokia 6600 phone. Then choose Software ⇨ More Software ⇨ Bluetooth Printing Application.

1. **Make sure that the printer is turned on.**

2. **Download and install the application using the steps in Technique 72.**

3. **When installation is complete, go to the phone's main menu and select the Print application (see Figure 9-38).**

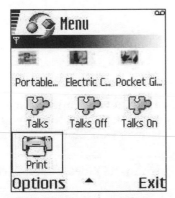

Figure 9-38: Select Print.

4. **Navigate through the folders on your phone to the image, message, contact record, or calendar entry (see Figure 9-39) that you want to print.**

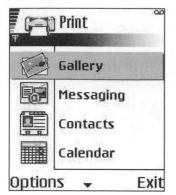

Figure 9-39: Navigate to the item you want to print.

5. **Select the image or file.**

   The print application will place a checkmark next to the item.

6. **Press Options on the left menu key.**

7. **Choose Print (see Figure 9-40).**

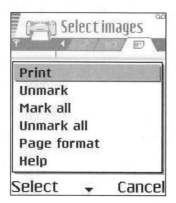

Figure 9-40: Choose Option ⇨ Print.

Now, you'll see the Select Printer screen.

8. **Select your printer from the list or, if you've never printed to a Bluetooth printer before, choose Option ⇨ Find Printer to ask your phone to perform a Bluetooth device discovery on Bluetooth printers within range.**

   Your selected image or file will then be sent to the Bluetooth-enabled printer.

Technique 54: Sending a File from Your PDA to Your Smartphone

My name is Dean Andrews and I'm a gadget head. I know my first step to recovery is admitting there is a problem. So, here I am in front of, hopefully, millions of readers, confessing my addiction to devices.

In my life, I've purchased clunky PIM devices, one of the earliest digital cameras, a variety of portable music players, voice recorders, data watches, a cellphone the size of a college dictionary, and other wacky gear. Some of the gear was useful — for a while — but eventually all these gadgets ended up collecting dust.

I found one way to get some value out of these devices was to remove my favorite or most important files from them before I put them on the shelf. Now, this is one of the first skills I learn when I buy a new gadget — how to move files from the device to somewhere else safe like a PC . . . or my smartphone.

The last device I purchased before I got my first Series 60–based phone was an iPAQ PocketPC. Like the other devices, it served my purposes for a while. But since I've learned all these cool smartphone techniques, I no longer want to carry a phone and a PDA. So, I'm transferring my most important files from my iPAQ to my smartphone.

Here's how to do it — just in case you have a little gadget problem of your own:

1. **On the iPAQ, press Start.**
2. **Select Programs.**
3. **Select File Explorer.**
4. **Navigate to the file you want to transfer to your smartphone.**
5. **Press and hold the stylus on the particular filename.**

   You'll see a pop-up menu appear.
6. **Choose Beam File.**
7. **On your phone's main menu, choose Connectivity.**
8. **Select Infrared.**

   Even the earliest iPAQ devices featured IR ports.
9. **Place your phone's IR port 1 to 2 inches away (facing) your iPAQ's IR port.**

   Your phone's name will appear on the iPAQ's Infrared Send screen.

10. **Select your phone from the list using the iPAQ's stylus.**

Note If your iPAQ comes equipped with Bluetooth, you can also use that to transfer files between your iPAQ and smartphone.

You'll see a progress screen showing the transfer from the iPAQ to the smartphone. The file will come attached to a message on your phone.

11. **On the main menu, select Messaging.**

12. **Select Inbox.**

13. **Select the message containing the file.**

14. **Press Options on the left menu key.**

15. **Choose Objects.**

16. **Highlight the object using the joystick.**

17. **Press Options.**

18. **Choose Save.**

19. **Select the area—phone memory or memory card—where you want to store the file.**

Repeat this process as necessary. Before you know it, all your most important information will be safely tucked away on your smartphone.

Technique 55: Using Your Smartphone as a Modem for Your PDA or Notebook

Have you seen cellular modem cards for PCs and PDAs? They can cost anywhere from $50 to $200, depending on the features and other types of connections—Bluetooth, Ethernet, and so on—that they provide.

As with USB storage devices (see Technique 33) though, you really don't need one. Your smartphone can do everything that a cellular modem card can. Your smartphone can act as a modem for your PC or PDA by establishing a Bluetooth or IR connection between your phone and your computer.

Tip Ask your service provider about software to make your smartphone work as a modem for your PC. Many service providers now supply very friendly software applications that take all the pain out of this task.

Follow these steps to use your smartphone as a modem for a PC:

1. **On your Windows PC, choose Start ⇨ Settings ⇨ Network Connections.**

2. **Click Create a New Connection.**

3. **Launch the Network Connection Wizard.**

4. **Follow the instructions in the wizard.**

   You'll need to provide:

   - Your Internet service provider access phone number
   - Your ISP username
   - Your password
   - Your domain
   - Either Generic IrDA modem to use an Infrared connection between your phone and PC, or Generic Bluetooth modem for a Bluetooth connection
   - A new name for this new connection

After you've created a new connection, you can double-check on the connection to start it.

 Note Also, double-check your phone's PC software. Some of these packages feature user-friendly wizards that walk you through the phone-modem setup procedure.

Securing Your Phone and Data

Security remains high on everyone's list of priorities — in most cases, with good reason. These days, it seems, not only are our computers vulnerable to attack, but even our identities can be stolen. It's hard not to be paranoid.

The bad news, in terms of your smartphone and security, is that a few phone viruses have now been reported though, fortunately, they've only been created in deep, dark, software-testing labs for research purposes and none have been found "in the wild." Still, this news chills the blood. And, of course, cellphones themselves have been reported stolen and misused frequently over the years.

The good news: Your smartphone features several built-in security settings, which allow you to monitor the use of your phone, restrict the use of your phone, and protect your phone against unwanted access to or theft of data — and your SIM card "identity." I walk you through all these techniques in this chapter.

Additionally, application developers are working hard in the area of cellphone security. If this topic is important to you, browse the Web resources mentioned throughout this book for VPN clients and antivirus software for your smartphone. Some is available now, and more is on the way.

Technique 56: Browsing and Erasing Your Call Log

Did you know your phone tracks every call it makes and receives? It also separately tracks the incoming calls that you answer as well as the incoming calls that you don't, totals the number of minutes you've used your phone for calling, and even tracks the number of kilobytes you've sent and received while online.

The application that does all this is called the Log (Records on Siemens SX1). In terms of security, the Log is great. By browsing the Log, you'll know in seconds if someone other than you has used your phone to make or receive a call or browse the Web.

Tip In standby mode, simply press and hold the Send key on your Series 60 phone. The phone automatically launches the Recent Calls section of the phone's Log application.

Follow these steps to browse your phone's Log:

1. **On the main menu, select Log (Records on the Siemens SX1).**

   On the Log's menu, you'll see Recent Calls, Call Timers, and GPRS Counter (see Figure 10-1). By browsing Recent Calls, Call Timers, and GPRS Counter, you can quickly tell how your phone is being used or if there is any unwanted usage of your phone.

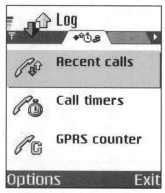

Figure 10-1: The Log's menu

2. **While you're here, press the joystick to the right.**

   You'll see the individual Log entries (see Figure 10-2) that are collected and categorized into the Recent Calls, Call Timers, and GPRS Counter categories.

   These individual entries (see Figure 10-3) are the "raw" data used by the Log.

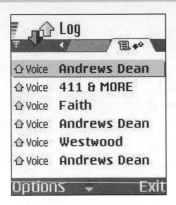

Figure 10-2: The individual Log entries of incoming and outgoing calls, SMS and MMS messages, and Web transactions.

Figure 10-3: An individual Log of a Web (also known as GPRS) session. Note that the date and time, duration, and amount of data is logged.

3. **Select an individual Log.**

   Note that it contains the type of transactions (call, SMS/MMS, or Web) the phone number involved, the length of the transaction, and it's date and time.

Tip

> You can sort, or *filter,* the list of Log entries. To do this, open the Log (Records on Siemens SX1) application. From the main screen, click right on the joystick. You'll see a list of all transactions. Then choose Options ➪ Filter while viewing the list. Your filter choices are: All Communication (No Sorting), Outgoing Communication, Incoming Communication, Voice Calls, Messages, Data Calls, and Selected Number. Selected Number is a special choice that allows you to quickly check how many times you or someone else called a particular number. Using the number that was highlighted when you chose Filter, the Log sorts to show you every entry of that number. Wow!

4. **Press Back on the right softkey to go back to the list of individual Log entries.**

5. **Press the joystick to the left to go back to the Log's main menu.**

6. Select Recent Calls.

Here's you'll see the categories: Missed Calls, Received Calls, and Dialed Numbers (see Figure 10-4).

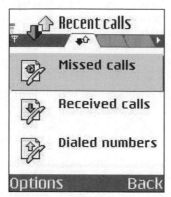

Figure 10-4: See all the calls you made, received, and didn't catch.

7. Select Missed Calls.

Here you'll see a list of all the calls you didn't answer (see Figure 10-5).

Figure 10-5: The list of calls you didn't answer. It's faster than checking voicemail. And it shows callers even if they left no message.

Tip

Make use of the Log's phone-number tracking for adding new entries to your Contacts directory. While viewing one of the three (Missed Calls, Received Calls, Dialed Numbers) lists, choose Options ➪ Add to Contacts. You'll be asked whether you want to create a new contact or add the number to an existing one.

Tip

I show you how to copy and paste information on your phone in Technique 26. The Log is a good resource for getting numbers that you might find handy for copying. Say you need to pass along a person's number to someone else in an SMS. While viewing one of the three (Missed Calls, Received Calls, Dialed Numbers) lists, choose Options ➪ Use Number. You'll be asked whether you want to edit or copy the number. To pass along the number in an SMS message, choose Copy. Then in the SMS edit screen, press the edit key (pencil) and choose Paste. Alternatively, you could choose Edit to use the Copy feature to highlight a specific portion of the number.

Tip

Want to start fresh? While viewing the Log's list, choose Options ➪ Clear List. All the entries will be removed and the logging will start again from the beginning.

I often use my Log just as a reminder of my day at work and with whom I've spoken. I also frequently use it as a source for adding new contact phone numbers to my directory. You can, too!

Keeping track of things

Someone else — maybe a teenage son or daughter — using your phone with your permission? You can adjust the Log to record for more or fewer (or no) days. Here's how:

1. **On the Log's main menu, choose Options ➪ Settings.**

   You'll see two parameters: Log Duration and Show Call Duration.

2. **Select Log Duration.**

   You'll see a list of choices: No Log, 1 day, 10 days, or 30 days (see the figure).

3. **Select one.**

Choose how many days you want your phone to track its usage.

4. **Press Back on the right menu key to return the Log application's main menu.**

Technique 57: Setting Restrictions on Your Phone

Remember I've said that Series 60 has a lot of cool features? Here's another one: Call Restrictions. It's available in all Series 60 phones and can be extremely handy in the right situation.

Note See Technique 49 for information on upgrading the software on your Sendo X.

The name of the application Call Restrictions is self-explanatory. You can set password-protected restrictions on how your phone can be used, and then hand your phone to someone you (mostly) trust. This means you can block incoming calls, outgoing calls, or both. You can block international calls or set the phone to allow only international calls that are calling the phone's home country. To save on phone bills, you can even block incoming calls only while roaming.

Note Call restrictions require four-digit passwords from your service provider. Check your service provider's Web site or call its technical support and ask about call-restriction passwords.

Here's what's going on under the covers: When you enter these call-restriction passwords on your phone in the steps that follow, your SIM card will be automatically placed on restriction lists inside your service provider's network.

Follow these steps to set restrictions on how your phone can be used:

1. **On the main menu, choose Tools ➪ Call Divert (Settings ➪ Call Restrictions on the Nokia 6260, Nokia 6620, Nokia 6630, Nokia 7610, Panasonic X700, and Samsung SGH-D710; Setup ➪ Call Divert on the Siemens SX1), as shown in Figure 10-6.**

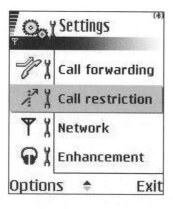

Figure 10-6: Select Call Restriction on your Settings menu. This is an example using the Nokia 7610 menus. Your menus may be slightly different, as mentioned in Step 1.

2. **Make your selection(s) from the list of restrictions (see Figure 10-7).**

   You can block all:

   - Outgoing calls
   - Incoming calls
   - Incoming calls while roaming
   - International calls
   - International calls except those to the phone's home country

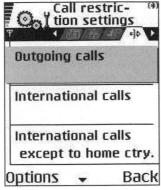

Figure 10-7: Make your call-restriction selections.

 Tip

Forgotten which restrictions you've already made? Choose Options ⇨ Check Status inside your Call Restrictions application. You'll be notified which call restrictions are active.

 Tip

Cancel all restrictions by choosing Options ⇨ Cancel All Restrictions.

3. **On the pop-up menu, select Activate (see Figure 10-8).**

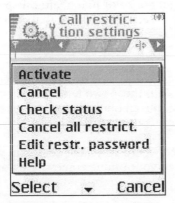

Figure 10-8: Select Activate on the pop-up menu.

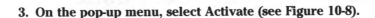

4. **Enter the restriction password you received from your service provider (see Figure 10-9).**

Figure 10-9: Enter the password you received from your service provider.

5. **Press OK on the left menu key.**

 Tip   Want more restriction options? Turn to Technique 60.

You'll receive a confirmation of your new restriction.

6. **Repeat the step for each new restriction.**

To remove an individual restriction repeat the steps, but choose Cancel instead of Activate.

Technique 58: Locking Your Phone

Lots of people have screensaver passwords on their PCs — if you're in this group, when you leave your computer to go get coffee or lunch, after a specified wait, your PC automatically throws up a screensaver picture that requires someone sitting back down at the keyboard to enter a password before that person can get back to the Windows desktop and get to work.

Did you know your cellphone has the same capability? You invoke it by setting a Phone Lock Code and an AutoLock Period. With these two features, your phone will lock automatically after a customizable amount of time. Anyone who wants to use your phone must first enter a code to unlock it.

When or where would this be useful? Anyplace where you must set down your phone in a public place, and anytime you place your phone somewhere you aren't.

Follow these steps to activate a phone lock on your phone:

1. On the main menu, choose Settings (Tools ⇨ Settings on the Nokia 7650, Nokia 3600/3620/3650/3660, Nokia N-Gage, Nokia N-Gage QD, and Sendo X; Setup ⇨ Settings on the Siemens SX1).

2. In Settings, select Security (Call Barring on the Sendo X).

3. In Security, select Phone and SIM.

4. Select Lock Code (see Figure 10-10).

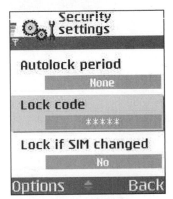

Figure 10-10: Select Lock Code so that you set a new lock code.

5. Enter the current lock code.

   Try the factory default if you've never set it (see Figure 10-11).

 The default factory lock code for a Series 60–based phone is 12345.

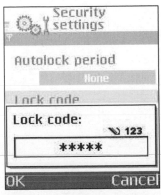

Figure 10-11: Enter your lock code.

6. When prompted to enter your new lock code, do so.

   Remember: Make it memorable.

7. When prompted to verify the new lock code, do so.

8. Now that you've changed the lock code to something familiar to you, scroll to AutoLock Period (see Figure 10-12).

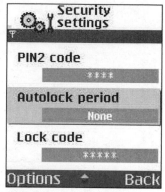

Figure 10-12: Scroll to AutoLock Period.

9. Select AutoLock Period.

10. Choose User Defined (see Figure 10-13).

Figure 10-13: Choose User Defined AutoLock Period.

11. In the Lock After pop-up, enter the number of minutes of inactivity after which you want the phone to automatically lock itself (see Figure 10-14).

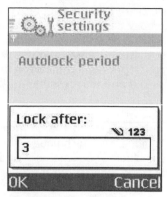

Figure 10-14: Enter the number of minutes for your timeout.

12. Press OK on the left menu key.

13. In the Lock Code pop-up, enter your new lock code.

14. Press OK on the left menu key.

You'll see a `Code Accepted` confirmation message.

Now after your specified AutoLock Period, you'll see the Lock Code pop-up when you press any key. Enter your lock code to enable use of the phone (see Figure 10-15).

Figure 10-15: When you press any key on your locked phone, you'll see the Lock Code pop-up.

Technique 59: Locking Memory Card Passwords

How valuable is the data and information you've stored on your memory card? If your answer is "Very," you'll want to set a memory card password.

A password protects your memory card against theft and unwanted use. If your memory card is ever removed from your smartphone and inserted into another device, the memory card will remain locked and inaccessible until someone enters the password.

 Note A memory card password does not protect access to the memory card while it remains inserted in your smartphone. Use Technique 58 to stop unwanted access to your whole phone, including your memory card. This memory card password technique only stops others from removing your memory card and being able to use it in another device.

Follow these steps to set a password for your memory card (except Nokia 7650, which does not include support for memory cards):

1. **On the phone's main menu, select Extras.**

2. **Select Memory (see Figure 10-16).**

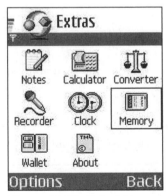

Figure 10-16: Select Memory in the Extras folder.

In the Memory Card menu, you'll see the overall capacity of your memory card, how much space has been used, and how much remains free (see Figure 10-17).

Figure 10-17: The basic stats of your memory card.

3. **Press Options on the left menu key.**

4. **Select Change Password.**

5. **Enter the current password, if one exists.**

6. **Enter the new password.**

7. **Verify the new password by entering it again.**

8. **Press OK on the left menu key to complete the change.**

 Tip   To remove a memory card password, choose Options ⇨ Remove Password. You'll need to enter the current memory card password to complete the change.

Now your memory card is password-protected against theft and unwanted use in another device.

Technique 60: Protecting Your Phone against a New SIM

Sometimes thieves don't want your SIM card (see "Terms you need to know" in the Introduction if you need a definition); they're just after your expensive phone. Their plan would be to steal your phone, open the case, throw your SIM card away, and sell the hot merchandise to someone else for use with a new SIM card.

 Tip   If your phone is missing or stolen, immediately call your service provider. They can deactivate your SIM card remotely so that it can't be useful in any phone.

By swapping in a new SIM card, thieves could *try* to make use of your smart-phone. Deactivating your SIM card doesn't help in this scenario, because your SIM card isn't being used, just your phone. But, in this technique, I describe how to foil this misuse of your phone.

Follow these steps to lock your phone against new SIMs:

1. **On the main menu, choose Settings (Setup on the Siemens SX1; Tools ➪ Settings on the Nokia 7650, Nokia 3600/3620/3650/3660, Nokia N-Gage, Nokia N-Gage QD, and Sendo X).**

2. **On the Settings menu, select Security (Call Barring on the Sendo X), as shown in Figure 10-18.**

Figure 10-18: Choose Settings ➪ Security.

3. **In Security, choose Phone and SIM (see Figure 10-19).**

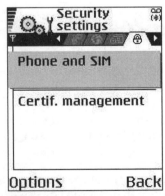

Figure 10-19: Choose Phone and SIM.

4. **Select Lock if SIM Changed.**

5. **Select Yes on the Yes/No screen.**

6. **Enter the lock code.**

7. **Press OK on the left menu key.**

Now you've protected your phone against use with an "unauthorized" SIM. Of course, this protection works best as a deterrent if potential thieves are aware that you've performed this technique. Perhaps someone will invent a phone sticker like those you see on cars: "This phone protected against theft by *101 Cool Smartphone Techniques!*"

Restricting dialing to only certain numbers

Under Phone and SIM security settings lies another powerful phone control feature — Fixed Dialing. This one is pretty extreme, so I only recommend it for desperate situations. It limits phone dialing to an exclusive list of numbers you specify. Thus, a reckless teen, a senile adult, or someone else who has proven he can't be trusted with a normally functioning cellphone — yet, for some reason still needs one — might require you to use this technique.

Note: You'll need an Extend Services SIM password from your service provider to enable this setting.

Follow these steps:

1. In the Phone and SIM menu, select Fixed Dialing.

You'll see a list of Fixed Dialing Contacts (see the figure).

A blank list of Fixed Dialing Contacts.

Continued

Continued

2. **Choose Options ⇨ Add from Contacts or Add New Contact to create a list of numbers.**

   Note: Emergency numbers like 911 in the United States will be active whether you add them or not.

3. **Enter the Extended Services PIN number that you obtained from your service provider.**

4. **Complete your list.**

5. **Press Done on the right menu key.**

6. **Choose Options ⇨ Activate Fixed Dialing (see the figure).**

Choose Options ⇨ Activate Fixed Dialing to limit the phone numbers your phone can call.

Technique 61: Encrypting Your Data Using the SmartCrypto Download

I'm sure you've heard: Bad things happen to good people. There's also the infamous Murphy's Law. In terms of phone security, these portents of doom imply that, at some point, your phone might be stolen and the thieves will have access to your personal information and all other data stored within it.

Tip   You'll certainly reduce your risk of thieves accessing your data by locking your phone with a lock code and locking your MMC with the other techniques in this chapter.

How can you protect your information? You can use a shareware download called SmartCrypto to encrypt the most important and private files resident on your phone.

Note What is encryption exactly? Instead of files being stored in plain text or some other readable format, encryption uses an alpha-numeric key as a formula to code (encrypt) a file into an unreadable format. To make the file readable again, the same key is used to reverse the formula on the file (or decrypt it).

SmartCrypto allows you to encrypt a set of files on your phone that you select using a key that you choose.

Note SmartCrypto can encrypt text-based files and image files.

Here's how to encrypt files using SmartCrypto:

1. **Download and install SmartCrypto from the SymbianWare Web site (www.symbianware.com) using the instructions in Technique 72.**

2. **On your phone's main menu, select SmartCrypto (see Figure 10-20).**

Figure 10-20: Launch SmartCrypto.

The first time you launch SmartCrypto, you'll see a blank list (see Figure 10-21).

Figure 10-21: SmartCrypto's
blank encryption list at startup.

3. **Press Options on the left menu key.**
4. **Choose Add Files to List (see Figure 10-22).**

Figure 10-22: Pick Add Files
to List to select the files
you want encrypted.

5. **Navigate through the folders and press the joystick down to select files to add to the list.**

Each click places a checkmark next to a file (see Figure 10-23).

Figure 10-23: Checkmarks appear next to the files you select.

6. **Press Add on the left menu key to add files into the list.**

   Now your SmartCrypto list shows the files you've added (see Figure 10-24).

Figure 10-24: The SmartCrypto list shows the files you added.

Tip
By default, SmartCrypto will remember the encryption key you use for each file. For extra security, change this setting. On SmartCrypto's main screen, choose Options ⇨ Session Options ⇨ Don't Keep Key.

7. **Highlight a file in the list.**

8. **Press Encrypt on the right menu key.**

9. **In the Password pop-up, enter your encryption key (see Figure 10-25).**

   Make it memorable, especially if you set the application to not remember the key.

Figure 10-25: Enter your encryption key.

10. **Press OK on the left menu key.**

11. **Verify the encryption key by re-entering it.**

12. **Press OK on the left menu key.**

   Now the file shows Encrypted File underneath it (see Figure 10-26).

Figure 10-26: Encrypted File appears underneath the files you've encrypted.

13. **Repeat steps 8 through 13 for every file in your list.**

After a file is encrypted, the right menu key becomes Decrypt when you highlight the file. You'll be asked to enter the encryption key and press OK. If you've set SmartCrypto to remember the key, all you need to do is press OK, because the encryption key will automatically appear in the pop-up window.

Technique 62: Protecting Your Phone from BlueJacking

What the heck is BlueJacking? It a way to anonymously send a message from one Bluetooth device to another Bluetooth device. Sounds harmless, doesn't it? Well, it's no threat to your data, but it might threaten your sanity.

BlueJacking, in the wrong hands, can be used to send you annoying messages that you don't want. For example, a friend and colleague recently spent a weekend in Las Vegas. In more than one casino his smartphone was bombarded with text-message ads from local businesses using the BlueJacking method. After a few frustrating minutes spent dismissing these bothersome messages one by one, he used the following technique to protect his phone against these advertisement "attacks."

 Note Because you must accept any Bluetooth connection attempted between your phone and another device, your phone is safe from a Bluetooth "hacking" type of attack on your phone's information and data. The deviously clever thing about BlueJacking is that it uses this security feature to deliver its message. Thus, when BlueJacked, your phone displays something like `"Joe's Pizza Palace delivers free!"` is attempting to make a Bluetooth connection to your phone. Accept or Reject?

 Tip There is only one sure way to combat BlueJacking and that is to set your Bluetooth communications to Off whenever you take your phone into an area that you can't trust.

Here's how to shut off your Bluetooth communications:

1. **On your main menu, choose Connectivity.**
2. **Select Bluetooth (see Figure 10-27).**

Figure 10-27: Select Bluetooth.

3. **Select Bluetooth again (see Figure 10-28).**

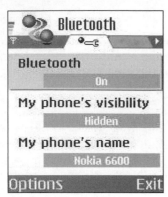

Figure 10-28: Select
Bluetooth again.

This action turns Bluetooth communications from On to Off (see
Figure 10-29). *Remember:* When you get back to your home or office,
turn your Bluetooth communication back on in order to transfer and
synchronize files over a Bluetooth connection.

Figure 10-29: Bluetooth
has been turned off.

Technique 63: Managing Digital Certificates on Your Phone

Digital certificates on your phone work just like digital certificates on your PC.
They're used when browsing the Web and when installing applications, to
establish trust between your smartphone and other entities. You need certificates, for example, when you interact with an online bank.

Some certificates come pre-installed on your phone when it arrives, others are downloaded from the Web when you download applications or browse the Web over your phone.

 Tip If you recently received your smartphone, it's a good idea to browse the certificates that were pre-installed on your phone. Because phone makers and service providers placed these certificates on your phone, they automatically have a higher level of trustworthiness than those you might end up downloading. Make a mental note of these pre-existing certificates so that you'll know later which were installed after you first received the phone. In case you later accidentally download certificates you didn't intend to, this mental note will better enable you to identify the unwanted certificates.

Follow these steps to view the digital certificates on your phone:

1. **On the main menu, select Settings (Tools ⇨ Settings on the Nokia 7650, Nokia 3600/3620/3650/3660, Nokia N-Gage, Nokia N-Gage QD, and Sendo X; Setup ⇨ Settings on the Siemens SX1).**

2. **In Settings, select Security (Call Barring on the Sendo X).**

3. **Select Certificate Management.**

   Here you'll see the list of digital certificates installed on your phone (see Figure 10-30).

Figure 10-30: The list of digital certificates on your phone.

 Tip To view personal certificates that have been granted to you, press the joystick to the right.

4. **To view the details of a certificate, particularly the date when it expires, select the certificate (see Figure 10-31).**

Figure 10-31: View certificate details like the validity dates by selecting the certificate.

5. **To view how a particular certificate is used (for the Internet or with applications, for example), choose Options ⇨ Trust Settings.**

   Now you'll see the settings showing in what circumstances the certificate will be used (see Figure 10-32).

Figure 10-32: Choose Options ⇨ Trust Settings to see how a certificate is used.

Tip

If you find a certificate that you don't trust, choose Options ⇨ Delete to remove it. You can also mark several certificates (using Options ⇨ Mark/Unmark) to delete more than one at the same time.

Managing Your Phone's Resources

At this point you probably realize that your smartphone is more like a computer than an old cellphone. With that sophistication comes the power — and sometimes the need — to manage your phone's resources a little like you adjust settings on a Windows PC. At work, you might have an IT system administrator care for your PC. Just try finding someone to care for your phone.

Because you're your own phone system administrator, in this chapter I show you how manage your phone memory, how to copy to and from your SIM card, and how to use phone utilities like the Application Manager and the Connection Manager.

Also in this chapter I show you how to make backups, format a memory card, and install applications. Take care of your phone's resources, and your phone will take care of you.

Technique 64: Viewing and Managing Phone Memory

What types of files are consuming the most memory on your phone? Is it your audio, image, or video files? Is your contact database taking more room than your message inbox?

Don't worry if you don't know the answers to these questions. Unless you've had lots of experience with your phone, you won't know how to view and manage your memory the way this technique shows you.

Follow these steps to view the "big picture" and the details of the phone's memory and memory-card usage:

1. **On your phone's main menu, select Tools.**

2. **Select File Manager (see Figure 11-1).**

Figure 11-1: Inside the Tools folder, select File Manager.

 Note   Not all Series 60–based smartphones feature the File Manager application. Read your phone's user manual if you're not sure whether your phone does.

Here you'll see phone memory on the left tab and your memory card on the right tab.

3. **While viewing the phone memory tab, press Options on the left menu key.**

 Tip   Press the joystick to the right to view the Memory Card tab. Press the joystick to the left to view the Phone Memory tab.

4. **Now select Memory Details (see Figure 11-2).**

Figure 11-2: Select Memory Details to go deeper into your memory card usage.

At this point, you see a breakdown of memory usage based on file type (see Figure 11-3). Use this information to prune files in your memory.

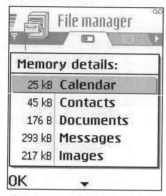

Figure 11-3: View the details of your memory card.

Repeat these steps, but in Step 3, press the joystick to the right to access your memory card. Then, follow the same steps to view your phone's memory card usage.

Clearing out space in your phone's memory

Use this technique to find out exactly what files are hogging your phone's memory. Then, take action! Here are some specifics based on what you find on your memory details screen:

✦ **Are image, audio, and video files consuming most of the space in memory?** If so, consider offloading those multimedia files to your PC (using Technique 52).

✦ **Are messages weighing down your phone memory?** Then, browse through your inbox and delete all unnecessary messages. Inside your Inbox choose Options ⇨ Mark/Unmark to mark either one or a whole set of files. Then, press the C (or Clear) key on your phone. You'll get a confirmation message asking if you want to delete the files you've marked.

✦ **Is your Calendar stuffed full with old appointments?** If so, go into your Calendar and choose Options ⇨ Delete Entry ⇨ Before Date to delete all the entries before a date you specify.

Technique 65: Copying Information to and from Your SIM Card

Do your have more than one phone? Are you planning to upgrade to a new phone? Do you like to travel light?

If you answered Yes to any of those questions, you need to learn how to move data onto and off of your SIM card. Keep in mind that even though there are means through which you can back up or transfer contact information, phone memory is not portable. If you switch phones, none of the contact information in your phone memory will go with you. Whatever contact information is stored on your SIM card will go with you, because the SIM card goes with you.

If you store your most important contact information on your SIM card (in addition to your phone memory), you maximize your portability. By keeping your current contact info on your SIM card, you can quickly switch phones by removing your SIM card from your current phone and inserting it into a new phone.

Here's how to view your SIM card directory:

1. **On your phone's main menu, select Contacts.**

2. **Press Options on the left menu key.**

3. **Choose SIM Directory (see Figure 11-4).**

Figure 11-4: Choose SIM Directory.

 Tip   Your SIM card holds a maximum of 254 contact records. To see how many you've used, choose Options ➪ SIM Details in the SIM Directory screen. You'll see the total number of contacts you have stored on the SIM card and how many free slots remain (see Figure 11-5).

Figure 11-5: Check how many of your SIM's 254 SIM contact slots you have filled.

Here you'll see the contact information resident on your SIM card (see Figure 11-6). If you've never copied contact information to your SIM, this list will be blank.

Figure 11-6: Your SIM card's directory listing.

4. **You can add a new individual contact to your SIM directory by choosing Options ⇨ New SIM Contact.**

Tip

To copy a contact from your SIM directory to your contacts directory, choose Options ⇨ Copy to Contacts from inside your SIM directory.

If you've added many contacts to your regular contacts directory, it's a good idea to copy them to your SIM directory for safekeeping. Here's how:

1. **On your phone's main menu, select Contacts.**

Note

SIM cards only store one number per name entry. The Series 60 Contacts application supports multiple numbers per name If you try to copy an entry that has multiple numbers to a SIM card, you'll be asked which entries you want to transfer. If you choose All, the numbers will be split into multiple individual entries and saved on the SIM with the same title (name).

2. **Highlight an individual contact.**

3. **Press Options on the left menu key.**

4. **Choose Copy to SIM Directory (see Figure 11-7).**

Figure 11-7: Copy selected contacts to your SIM directory for safekeeping.

Tip

You can copy multiple contacts simultaneously by using the contacts marking feature. Use Technique 28 to mark several contacts. Then, choose Options ⇨ Copy to SIM Directory to complete the maneuver. ***Note:*** This does not work on the Sendo X, Siemens SX1, Nokia 7650, Nokia 3600/3620/3650/3660, N-Gage, or N-Gage QD.

Technique 66: Making Backups of Your Phone's Data

Remember in this chapter's introduction when I said your smartphone was like a computer? Well, just as with your PC, it's a good idea to make regular backups of your information and data.

Wait! I see you running for the door, pulling your hair out. Don't worry! Very much unlike PCs, making a backup of your phone's memory is easy and fast, almost completely painless.

Follow these steps to make a backup of your phone's memory using your memory card. (***Note:*** The Nokia 7650 is the only Series 60–based phone that does not support external memory cards.).

1. **On the main menu, select Extras.**

2. **Select Memory (see Figure 11-8).**

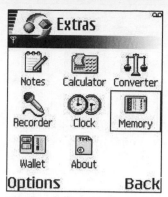

Figure 11-8: Select Memory inside the Extras folder.

3. **Inside the Memory Card application, press Options on the left menu key.**

4. **Choose Backup Phone Memory (see Figure 11-9).**

Figure 11-9: Choose Backup Phone Memory from the Options menu.

5. **At the confirmation message (see Figure 11-10), press Yes on the left menu key.**

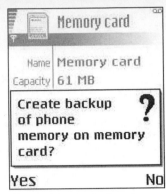

Figure 11-10: At the confirmation message, press Yes.

You'll see a progress bar as the phone makes its own backup onto your memory card. Then, a `backup complete` message appears when the task completes.

> **Note**
>
> Any alternatives to making a backup on your memory card? Sure. You can use your PC software (see Techniques 49 through 52 on finding and using PC software for your phone). Most PC software packages offer a phone backup. Instead of creating a backup on your memory card, these utilities create backups to your PC and allow you to restore the data from there as well.

How do you restore your phone from a backup you've made to your memory card? Inside the Memory Card application, choose Options ⇨ Restore from Card (see Figure 11-11). You'll be asked to confirm your request. Press Yes on the left menu key. And—*voilà!*—your wish is the phone's command.

I told you it was easy. If you make a backup to your memory card every week or two, you'll be completely protected against any kind of disaster—dropping your phone off a cliff, the latest hurricane to pass through Florida, or whatever. Even if your phone is damaged, your memory card should survive just about anything.

Figure 11-11: Restore your data using your backup.

Technique 67: Formatting New Memory Cards

Bigger is better when it comes to multimedia memory cards. Every few months, you'll see prices drop and you'll start to wonder if you need to upgrade the memory card in your phone.

If you do purchase a new memory card, you may need to format it. Here's how:

 Note   Read the instructions that come with your new memory card. In some cases memory cards come preformatted, and you won't have to perform this technique to make the memory card work in your phone.

1. **On the main menu, select Extras.**

2. **In the Extras folder, select Memory.**

3. **Inside the Memory Card application, press Options on the left menu key.**

4. **Choose Format Memory Card (see Figure 11-12).**

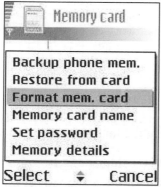

Figure 11-12: Format a new memory card quickly and easily.

You'll see a warning message Format data card? Data will be deleted during formatting (see Figure 11-13).

 Caution   No kidding! Be careful. The format utility isn't smart enough to know if you're being stupid. If you decide to format a memory card that holds your family images, your favorites videos, and all your most important documents, it will let you. Everything on the memory card will be destroyed during formatting.

5. **Press Yes on the left menu key only when you're sure about what you're doing.**

Figure 11-13: Think before you format a memory card.

Technique 68: Saving Memory by Adjusting Your Log Settings

Did viewing your phone's memory in Technique 64 show you that you don't have that much free phone memory? In that technique, I describe the steps you need to take to reduce the space consumed by multimedia files, contacts, and messages, but there is one more space consumer that deserves its own technique—your Log application.

By default, the Log application runs all the time, taking note of every phone call you make or take, every SMS/MMS you receive or send. If phone security is a big concern, then use and configure the Log application as I describe in Chapter 10. If, on the other hand, you're less worried about security and more interested in saving space and memory, perform the following steps:

1. **On the main menu, select Log (Records on Siemens SX1).**

2. **On the Log application main screen (see Figure 11-14), press the joystick to the right.**

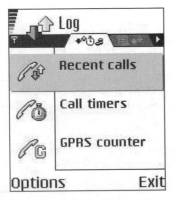

Figure 11-14: The Log application's main screen

Here you see the raw data log (see Figure 11-15).

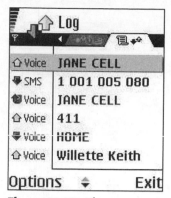

Figure 11-15: The raw data log

3. **Press Options on the left menu key.**

4. **Choose Clear Log (Clear Records on Siemens SX1), as shown in Figure 11-16.**

Figure 11-16: Choose Clear Log to clear out the raw data log of the Log application and save some space.

You'll see a confirmation message asking `Clear all communication details from log?`

5. **Press Yes on the left menu key.**

Now, that you cleared out the Log let's adjust the settings so that your phone does not create a new log.

6. **Press Options on the left menu key.**

7. **Choose Settings.**

8. **Select Log Duration (Records Duration on Siemens SX1).**

9. **Select No Log (see Figure 11-17).**

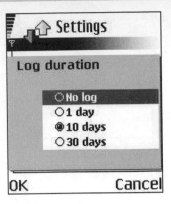

Figure 11-17: Select the No Log setting to save more space in your phone's memory.

You'll see a confirmation message telling you Log, recent calls and delivery receipts will be deleted and not registered. Continue?

10. **Press Yes on the left menu key to confirm.**

Congratulations! You're officially a super phone IT system administrator!

Technique 69: Making Smarter Image and Video Settings

Are you taking advantage of your phone's digital camera and video camcorder capabilities? The image and video features of these smartphones are some of the most exciting of all. But the output of the camera and camcorder also consume more space than any other application.

By default, both the camera and the video camcorder applications place their images and video into the phone memory, not the multimedia memory card. This happens even though memory cards offer much more storage capacity than your phone memory. If you have a memory card, it's much smarter to change the settings of the camera and video so that they use the memory card for storage. Here's how:

1. **On the main menu, select Camera (Snapshot on Siemens SX1).**

2. **Inside the camera application, press Options on the left menu key.**

3. **Choose Settings.**

    On the Nokia 6620, 6630, 6260, 7610 and Samsung SGH-D710 *only*, you'll see the settings for both Image (digital camera) and Video (video camcorder), as shown in Figure 11-18.

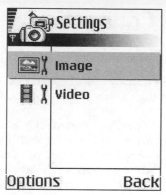

Figure 11-18: The Image and
Video settings of your
smartphone

4. **Select Image.**

5. **Inside Image Settings, select Memory in Use.**

   The setting will change to Memory Card.

   On many newer Series 60 phones, you can change the memory in use
   for the Camcorder application. For the Nokia 6620, 6630, 6260, 7610,
   and Samsung SGH-D710, follow the following steps after the preceding list:

6. **Press Back on the right softkey.**

7. **Select Video.**

8. **Choose Settings.**

9. **Select Memory in Use just as you did earlier, to change the setting
   to Memory Card.**

Tip   This tip only applies to Nokia 6620, 6630, 6260, 7610, and Samsung SGH-
D710. Now, if you want, you can change the Length setting on this same
screen to maximum. You won't be consuming your phone's memory, just
some of the free space on your big memory card.

Great job! Now every picture you snap — and, if you have a late-model Series
60–based phone, video you capture — will be dropped automatically onto your
memory card. If you have a memory card reader on your PC, you can even use
it to move images and video from your phone to your PC. Just pull it out from
the phone and insert it into your PC's memory card reader.

Technique 70: Using the Connection Manager

Any crackerjack IT computer administrator can quickly tell whether a PC is connected to a Web server and can also check the details of that connection.

Guess what? On many newer Series 60–based phones, you can quickly and easily check connections between your phone and Web servers using a utility called the Connection Manager. Through the Connection Manager, you can see if there are any connections, whether they are active or inactive, if data is being transferred back and forth, and more. You can also cancel all connections, with the click of one key. Here's how:

The Connection Manager application can only be found on the Nokia 6260, 6600, 6620, 6630, and 7610, Panasonic X700, and Samsung SGH-D710.

1. **On the main menu, choose Connectivity.**

2. **Select Connection Manager (see Figure 11-19).**

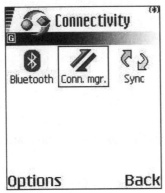

Figure 11-19: Select the Connection Manager utility.

Inside the Connection Manager, you'll see a list of all current connections your phone has with computer servers (see Figure 11-20).

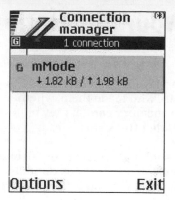

Figure 11-20: View the list of current connections between your phone and the world.

3. **To view the details of any connection, select it.**

 You can also choose Options ➪ Details to view a connection's details.

On the details screen, you'll see the amount of data sent and received, the duration of the connection, and more.

4. **Press OK on the left menu key to return to the connections list.**

5. **To cancel any connection, highlight it and press the C (or Clear) key.**

You'll see a confirmation message (see Figure 11-21).

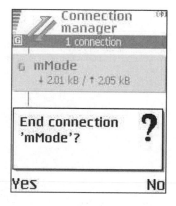

Figure 11-21: To cancel a connection, highlight the connection and press the C key.

It's nice to have access to the right tools and know how to use them. Now you do.

Technique 71: Viewing Your Application Manager's Log

Have you kept track of all the applications and games you downloaded to your phone? Do you know where they all came from, who made them, whether they're stored in your phone memory or your memory card, if they're Symbian-based or Java applications? Yeah, I didn't think so.

Fortunately, your phone has kept track of these things. Your Series 60–based phone has a cool utility called the Application Manager that has dutifully tracked every application you installed on your phone. And it knows more about each one than you know about what's currently in your own wallet.

 Note This feature is not available on the Nokia 7650.

Follow these steps to use the Application Manager:

1. **On the main menu, select Tools (on Siemens SX1, choose Setup ⇨ Manager).**

2. **Inside the Tools folder, select Application Manager (see Figure 11-22).**

Figure 11-22: Select the Application Manager.

Here you'll see a list of all the applications you've installed on your phone (see Figure 11-23).

3. **To view the details, select an application.**

   On the details list, you'll see whether the application is Java- or Symbian-based, where it's stored, who made it, how big it is, and more.

4. **Press OK to return to the main screen.**

Figure 11-23: The Application Manager shows a list of every application installed on your phone.

If you decide you need to remove an application from your phone, you should do that from the Application Manager as well. Simply highlight the application in the list and choose Options ➪ Remove. You'll see a confirmation message asking you if you really want to remove the application from your phone. Press Yes on the left menu key to confirm.

Related to the preceding tip: You may notice some applications feature a Delete choice on their Options menu. Choosing Delete will not properly "uninstall" the application. To properly "uninstall" use the Application Manager.

The Application Manager's log

There's one more feature to the Application Manager—the log file. The Application Manager contains its own log that records the name and date of every application you've installed and any errors that occurred during an installation.

You can view the log by choosing Options ➪ View Log inside the Application Manager.

If you're talking with a technical support representative of your service provider or an application maker, you can send him the log from your phone quickly and easily by choosing Options ➪ Send Log ➪ Via Text Message. The log will be automatically loaded into an SMS message, and you'll be transported to the message editor, where you can enter the recipient's phone number and send the message.

Technique 72: Installing Symbian Software

Many techniques in this book describe Symbian applications that you can download from the Web. Fortunately, there are fantastic and creative application developers out there who produce applications that are useful and fun. Here's how to install them:

1. **Download the application to your PC following the instructions on the Web site.**

2. **On your PC, launch your PC Suite software (in this example, I use Nokia's PC Suite Version 6).**

3. **On the PC software's main menu, select Install Applications (see Figure 11-24).**

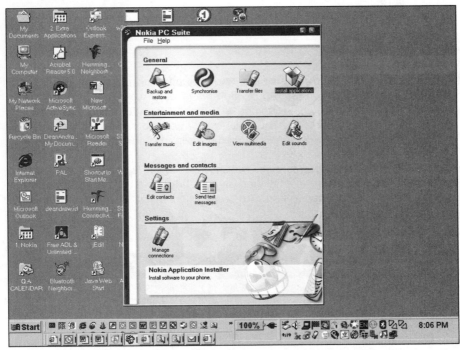

Figure 11-24: Select Install Applications on your PC software's main menu.

4. **On the My Computer (left-hand side) section of the Application Installer, navigate to the folder that contains the Symbian application (SIS file) you would like to install on your phone (see Figure 11-25).**

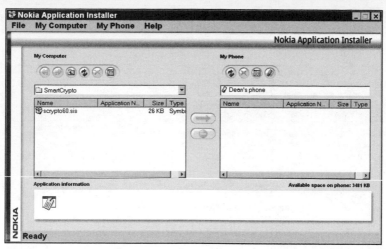

Figure 11-25: Navigate to the application file you would like to install on your phone.

Note All Symbian applications have an SIS file type.

5. **Press the arrow (or transfer) button to begin the installation (see Figure 11-26).**

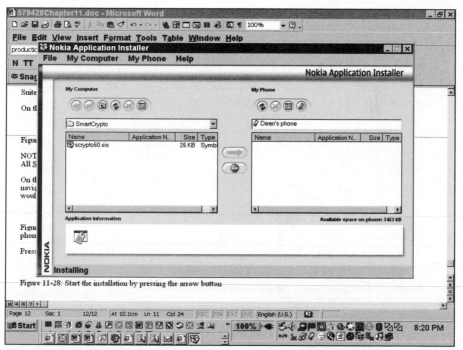

Figure 11-26: Start the installation by pressing the arrow button.

In most cases you see a Window asking you to complete the installation on your phone (see Figure 11-27).

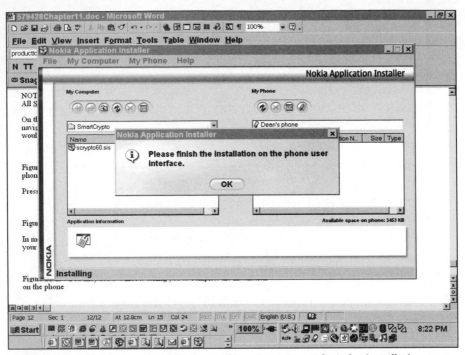

Figure 11-27: You may see a Window asking you to complete the installation on the phone.

On the phone, you'll see a screen asking if you trust the application provider (see Figure 11-28). If you downloaded one of the applications mentioned in this book, you can trust the provider.

Figure 11-28: Press Yes if you trust the application provider.

6. **Press Yes on the left menu key**

   Next you'll see a screen asking if you want to install the application (see Figure 11-29).

Figure 11-29: Press Yes to continue the installation.

7. **Press Yes.**

   You'll see a pop-up Options menu.

8. **Select Install and press Yes to continue (see Figure 11-30).**

Figure 11-30: Press Yes again to continue.

Next you'll be asked whether to install the application in phone memory or your memory card (see Figure 11-31). In most cases, the smartest choice is the memory card (more storage and it's not competing with other applications for limited space).

Figure 11-31: Phone memory or memory card? Usually memory card is the better choice.

9. **Highlight your choice, and press OK on the left menu key.**

10. **Press OK on the final information screen.**

11. **Back on your PC, click OK in the window that asks you to complete the installation on your phone.**

Tada! Not too bad was it? Install one or two Symbian applications, and you'll feel like an old pro using your PC software's application installer.

Tip

There is an alternative way to install a Symbian application on your smartphone. You can simply use Bluetooth or IR (or, for those phones that have USB, that as well) to connect to the PC. You can send the SIS file from wherever it is stored on your hard drive directly to the phone. It'll give a "1 New Message" indicator (the same one you receive with an SMS, for example) and the application will show up in your inbox. When you open the message, the installer will open and you'll be taken through the same installation sequence described earlier.

Cool Web sites for Symbian applications

Here are some cool Web sites for downloading Symbian applications:

+ **SymbianWare:** www.symbianware.com

+ **Download.com:** www.download.com

+ **My-Symbian.com:** www.my-symbian.com

+ **Forum Nokia:** www.forum.nokia.com

+ **Nokia Software Market:** http://softwaremarket.nokia.com

+ **Handango:** www.handango.com

Technique 73: Installing Java Software

Your phone runs both Symbian applications and Java applications. What's the difference between installing one versus the other? Mainly, the source for the applications and the file types of the files.

Note All Java applications have a JAR file type.

Follow the same procedure as described in the last technique, except in Step 3, navigate to the JAR file that contains the Java application you want to install on your phone.

Cool Web sites for Java applications

Here are some cool Web sites for downloading Java applications:

✦ PocketCinema: www.pocketcinema.com

✦ Download.com: www.download.com

✦ My-Symbian.com: www.my-symbian.com

✦ Nokia Software Market: http://softwaremarket.nokia.com

✦ Handango: www.handango.com

Mastering Messaging

No doubt about it. Messaging over phones is a worldwide phenomenon. Whether you're transacting business, staying in touch, conveying information, or flirting, a message from your phone can get the job done.

Thanks to the range of formats — text (SMS), multimedia, e-mail, or IM (instant message) — a message can be subtle or direct, clear or mysterious, dramatic or terse, polite or incredibly rude, and artful or utilitarian. Strangely enough, messages seem to allow for more freedom of expression than a regular voice call.

By this point, you've learned the basics of messaging. Here, I show you how to "take it to the next level." In this chapter, I show you a new way to quickly type a text message, how to turn a regular multimedia message into an impressive presentation, how to download and install an IM application, and more.

Technique 74: Writing Faster Text Messages Using Predictive Text

All Series 60–based phones offer a capability called predictive text input. With this feature turned on, the phone will try to "predict" or guess the word you're typing, rather than have you enter each character exactly.

 Note   T9 is the specific predictive text product used in Series 60–based phones. You may see predictive text referred to as "T9 input" in your phone's manual for this reason.

Normally, when entering a text message using a traditional phone keypad, you'd press each key several times to switch between letters. For example, you'd press the

2 key once to input the letter *A*, twice for the letter *B*, and three times for the letter *C*.

Predictive text input doesn't work like that at all. With predictive text you press a key only once, no matter which of the three alphabet characters you're trying to input. The predictive text algorithm predicts which word you're typing by comparing all the possible words in its dictionary that are made up of letters available on the keys you pressed. At the end of each word, you check whether the prediction was correct. If it is, you move on to the next word. If it's not, you correct it by choosing another word from its list of possible words.

The algorithm can "learn" more about your particular message writing style as you tell it to add words you use to its dictionary.

Is it really faster? It can be. For example, to input the word *how* without predictive text, you'd need to press six keys: the 4 key twice for *h*, the 6 key three times for *o*, and the 9 key once for *w*. With predictive text on, you only need to press three keys: the 4, 6, and 9 keys each once in order and — bang! — you'll see the word *how* on your phone's screen.

Here's how to use predictive text input:

1. **On the main menu, select the Extras folder (on Siemens SX1, select the Organizer folder).**

2. **Select Notes.**

 Note You can use predictive text wherever you enter text — while typing a new text message, creating a new contact, entering a new appointment in your calendar, adding a new note, and so on.

3. **Press Options on the left menu key.**

4. **Choose New Note.**

   In the upper-right-hand corner of the Note Editor, you'll see the letters *Abc* next to an icon that looks like a pencil (see Figure 12-1).

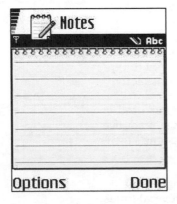

Figure 12-1: Notice the standard input mode symbol Pencil Icon Abc in the upper-right-hand corner.

5. **Press the Edit key (it looks like a pencil) on your phone.**

6. **Select Predictive Text (T9 or Dictionary in some phones) On (see Figure 12-2).**

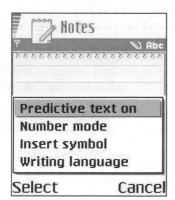

Figure 12-2: Select Predictive Text On.

You'll notice the symbol in the upper-right-hand corner change. Now there's a line running up to the pencil icon. This is the symbol for predictive text input activated.

7. **Enter the following two keys: 6 and 9.**

You should see the word *My* on the screen (see Figure 12-3).

Figure 12-3: Enter *My* by pressing the 6 and 9 keys once each.

8. **Now, examine the other "possible" words in the list by pressing the * key.**

   With each press, you'll see alternatives, which should include: *Oz, Ox, Ow, Ny,* and *Oy.*

9. **When you get back to *My,* press the 0 key (1 key on the Siemens SX1) to create a space between words and move on to the next word.**

 When the word you want is showing in underlined predictive text mode, you can also press the joystick to the right to select the word but without the space after it that the 0 key adds.

10. **Continue entering the phase *My first predictive text message* just to get used to the method.**

11. **Now, to add a word to the dictionary, enter the keys for *Wiley* (the publisher of this book), which would be 9, 5, 4, 3, and 9.**

    Instead of Wiley, you'll see wkge?

12. **Press Spell on the left menu key.**

13. **Enter Wiley into the Word field.**

14. **Press OK on the left menu key.**

    You'll see the misspelling replaced by the word you entered. And the next time you enter that key sequence, *Wiley* will be on the list of possible words.

You might find using predictive text unfamiliar and a little strange at first, but I recommend sticking with it for about a week of messaging. By the end of a week, you'll be able to write predictive text messages very quickly and smoothly.

Technique 75: Finding Information within a Message

I've showed you how to send virtual business cards. Sometimes, though, valuable information — like a phone number, an e-mail address, or a Web address — will arrive in a message, but the message won't be in the standard business-card format.

What to do? Go tell your phone to find the info you need! Here's how:

1. **On the main menu, select Messaging.**

2. **Select Inbox.**

3. **Scroll to a message you know contains a phone number, Web address, or e-mail address.**

4. **Select the message.**

   Now you're viewing the message.

5. **Press Options on the left menu key.**

6. **Scroll down until you highlight Find (see Figure 12-4).**

Figure 12-4: Look for Find on the message viewer's Options list.

7. **Select Find.**

   Now, you have a choice. For what information do you want your phone to search?

8. **Make your choice from the list: Phone Number, E-mail Address, or Web Address (see Figure 12-5).**

Figure 12-5: Tell your phone to search for a phone number, e-mail address, or Web address.

If successful, you'll see the information you requested highlighted and underlined in the message (see Figure 12-6).

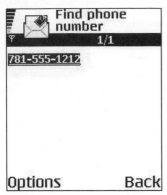

Figure 12-6: The phone will highlight and underline the information it finds.

What happens if the message doesn't contain any real information or the search is unsuccessful? A window pops up telling you the search was unsuccessful.

What can you do with information? Save it to your contact directory! And, the phone will help you do it.

9. **Press Options on the left menu key.**

10. **Select Add to Contacts (see Figure 12-7).**

Figure 12-7: Select Add to Contacts to save the information to your contacts directory.

Note   As an alternative, you can also select Copy from the Options menu and paste the information wherever you want.

Your choice is Create New or Update Existing (see Figure 12-8). This is simply asking whether you want to create a new contact record or modify one that is already present in your contact directory.

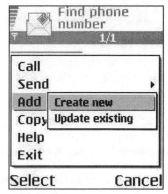

Figure 12-8: Choose either Create New or Update Existing.

11. Pick one of the options.

Now you'll be asked to select the detail — like mobile number, home number, e-mail address, Web site — into which you want to store the information (see Figure 12-9).

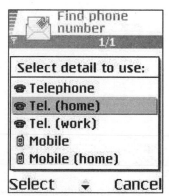

Figure 12-9: Select where you want the phone to store the information.

12. **Choose a field.**

    At this point, you'll be placed into the Contact Editor with the field you selected already filled in with the information from the message.

13. **Complete the entry of the contact record and press Done on the right menu key when finished.**

Technique 76: Changing the Writing Language of Your Smartphone

Your smartphone features two core customizable language settings: Phone Language and Writing Language.

Phone Language adjusts the menus, title bars, and options of almost every application on your phone. It also affects the time and date format used on the phone and separators used in calculations. Most Series 60 smartphones come pre-installed with three or four languages. You can find out which are pre-installed on your phone by choosing Settings ⇨ Phone Settings ⇨ General ⇨ Phone Language (see Figure 12-10).

Figure 12-10: View the different phone languages available on your phone.

 If you choose Automatic for Phone Language, your smartphone will use information stored on your SIM card to decide which language to use.

Writing Language adjusts the language used for creating messages, and that's the one I focus on here.

You can also modify Writing Language inside your phone Settings by choosing Settings ⇨ Phone Settings ⇨ General ⇨ Writing Language (see Figure 12-11).

Note   Settings is in the Tools folder on most Series 60 1st Edition phones — except for Siemens SX1. It's in the Setup folder on the SX1.

Figure 12-11: View the different writing languages available on your phone.

When you change the Writing Language setting, you're changing three things at once:

✦ **The characters you see when you press keys 1 through 9 at text entry fields like Notes, Calendar, Contacts, and during any message writing:** This means if you select Portuguese as the writing language, for example, you'll cycle through a different set of characters with repeated pressing of the 6 key than if the writing language were English.

✦ **The predictive text dictionary:** This means that, for example, if you select Spanish as the writing language, predictive text will predict Spanish words that match the key sequences you type, rather than English words.

✦ **The special character sets:** This means when you press the * key or the # key, you'll see different tables of symbols that you can insert into a message or text field across the different languages.

If you're fluent in more than one language, feel free to modify your writing language on the fly. You can do so by pressing the Edit key (it looks like a pencil on most Series 60 phones). Then select Writing Language (see Figure 12-12). You'll see the same list of languages you would see by viewing the phone settings. If you modify the writing language in either location, the change will be reflected in both locations.

Note   The Edit key only works when you have the cursor in a text entry field — like a contact's name field. If your cursor is on the main menu or anywhere else, pressing the Edit key does nothing.

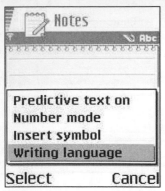

Figure 12-12: Select Writing Language from your Edit key's menu.

Technique 77: Creating a Multimedia Message That Looks Like a Slide Show

You know how to create text and multimedia messages. But you can actually create grander and more dramatic messages than that using your smartphone's Synchronized Media Integration Language (SMIL) capability.

 Note SMIL capability is only supported on Series 60 2nd Edition–based phones. Read the introduction of this book regarding this topic or check your phone's user manual if you're unsure whether your phone supports SMIL.

An SMIL message is like a slide show that can automatically flip between different images, have a soundtrack, and play a video. You can let your imagination run wild. It's the next best thing to making your own Hollywood production. And you can create it right on your phone and send it to anyone!

Here's an example:

1. **On the phone's main menu, select Messaging.**

2. **Select New Message.**

3. **Choose Multimedia Message (see Figure 12-13).**

   Here in the Message Editor notice what you may never have noticed before—the title at the top says "Multimedia Slide 1/1." When you created multimedia messages in the past, you probably didn't know or realize that you can add new slides, move them around, add sound, and more. The phone user manuals only gloss over this, if they mention it at all.

Figure 12-13: Choose Multimedia Message.

In this example, let's use this first slide as the title of our little slide-show "production."

4. **Enter** Shiny Happy People!

Tip

If you really want to get fancy and center the title on this "title page," insert a few "carriage returns" by pressing the * key to bring up the symbols table and selecting the return symbol (see Figure 12-14).

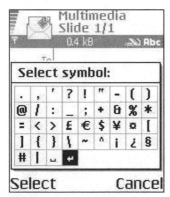

Figure 12-14: Select the return symbol to center the title.

Before we do anything else, let's add a soundtrack to our show. The sound will start playing when the slide that the clip is attached to is displayed, so if you attach the clip to the first slide, the soundtrack will play from the beginning.

5. **Choose Options ➪ Insert Object ➪ Sound Clip (see Figure 12-15).**

Figure 12-15: Insert a sound clip for your soundtrack.

6. **Navigate to a music clip or ring tone on your phone and select it.**

   Now, let's add a new slide to our slide show.

7. **Choose Options ➪ Insert New ➪ Slide (see Figure 12-16).**

Figure 12-16: Insert a new slide.

Let's insert an image on this slide, one of our "shiny happy people."

8. **Choose Options ➪ Insert Object ➪ Image.**

9. **Navigate to a folder and select an image.**

   It will appear on the slide. Notice the "Slide 2/2" on the title bar (see Figure 12-17).

Figure 12-17: Notice the "Slide 2/2" title. This is the second slide of your multimedia slideshow.

10. **Repeat steps 7 through 9 to add more slides with one image per slide.**

11. **When you've finished adding slides, press your joystick up repeatedly until you return to the first slide.**

12. **Add a recipient in the To field and send the message.**

 Tip
Choose Options ➪ Preview Message to see different slides of you slideshow. This isn't exactly the way your recipients will see it, however. For that, send the message to yourself!

A recipient of the message will open the message in his Inbox and choose Options ➪ Play Presentation (see Figure 12-18) in order to see your slide-show production.

Figure 12-18: Recipients will choose Options ➪ Play Presentation to see your slideshow.

Technique 78: Checking Your Message Size Limits

The more you experiment with multimedia messaging, the more you'll encounter the message size limits of your particular smartphone setup. By "setup" I mean your phone, your service provider, the mail servers with which you interact, and other network elements.

 Note Service providers have a size limit of 100KB for MMS messages. This is an agreed-upon MMS standard and is a main reason why large messages don't always go through.

You'll know when you've reached the size limits because either your message will never leave your phone or it may never reach your intended recipient (if you're lucky, your service provider will notify you of that fact via a text message or some other form of communication).

In this technique, I just want to show you how to determine the size of your messages and the size of the individual multimedia objects within them. If you run into problems with sending multimedia, try removing some of the largest multimedia objects or just creating a new, smaller message.

Here's how to check the size of a multimedia message and the multimedia objects inside it:

1. **On the main menu, select Messaging.**
2. **Select Inbox.**
3. **Select an individual multimedia message.**
4. **Press Options on the left menu key.**
5. **Scroll to highlight Message Details (see Figure 12-19) and select it.**

Figure 12-19: Select Message Details.

6. **Scroll down until you see Size.**

   It will be listed in kilobytes (KB).

7. **To view the size of individual multimedia objects, press OK on the left menu key to return to the message, and then choose Options ➪ Objects.**

   On the objects screen, you'll see a list of all the multimedia objects that make up the message and the size of each (see Figure 12-20).

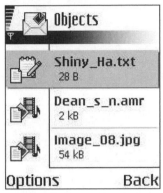

Figure 12-20: A list of all the multimedia objects in the message and their sizes.

Technique 79: Popping a Web-Based E-Mail Account to Your Phone

You can use your smartphone to send and receive Web-based e-mail while on the road. All it takes it a little setup and, depending on your Web-based e-mail service, a little investment. Yahoo! Mail (http://mail.yahoo.com), for example, requires you to sign up for its premium e-mail service—about $20 per year—before it will let you "pop" your Yahoo! Mail account.

Note The term *popping* comes from the POP3 protocol that standardizes a way for people to access mail servers and download their e-mails to a PC, a phone, or another mobile device.

The first step is to browse the tech-support or help section of your Web-based e-mail service and find out:

✦ Whether the service supports the POP3 protocol.

✦ Whether you'll incur any extra charge for popping your e-mail account.

✦ The specifics of the incoming and outgoing mail servers, including any specific port information. You'll need this during the e-mail setup on your phone.

After you have that information, here's how to set up the mailbox on your smartphone:

1. **On the main menu, select Messaging.**

2. **Select Mailbox.**

   The first time you select Mailbox, you'll be asked to set up a Mailbox profile.

3. **Select Yes on the left menu key.**

4. **Choose your Mailbox profile settings.**

   Here's a breakdown of the Mailbox profile settings and my advice for setting them:

   - **Mailbox name:** You can leave it as Mailbox. I only suggest changing it to something more specific if you're going to pop more than one account to your phone.

   - **Access point in use:** If you've already set up your multimedia messaging and Web browsing access points using Technique 2, just select this setting and choose your access point from the list.

   - **My e-mail address:** Enter your full e-mail account address like myaddress@isp.com.

   - **Outgoing mail server:** You'll need to get this info from your Web-based e-mail service.

   - **Send message:** I recommend setting this to During Next Connection, so that you can control when e-mail is sent and retrieved. The other option is Immediately.

   - **Send copy to self:** No.

   - **Include signature:** No.

   - **Username:** The name you use to log on to your e-mail service.

   - **Password:** The password you use to log on to your e-mail service.

   - **Incoming mail server:** You'll need to get this info from your Web-based e-mail service.

   - **Mailbox type:** POP3.

   - **Security:** No (or check with your service provider).

   - **APOP secure login:** No (or check with your service provider).

5. **Press Back on the right menu key when you've finalized the profile.**

   Now when you enter your Mailbox, you can create new messages and send and receive e-mail using the Options menu inside the Mailbox (see Figure 12-21).

Figure 12-21: The Options menu inside your Mailbox will let you create new e-mail messages, send them, and receive them.

Technique 80: Sending Instant Messages Using the IM+ Download

Instant messaging (IM) isn't just for fun anymore. Businesses have caught on to the amazing efficiency and immediacy of communication that IM provides. People who work in different states or even different countries can quickly collaborate to solve problems, come to agreement, or just exchange ideas through this powerful tool.

Oh, yeah, and it's still for fun as well.

Now you can take your chat with you when you go mobile. Some Series 60–based smartphones come with pre-installed IM or chat clients. Feel free to follow the instructions and use them.

If your phone didn't come with an IM client (or it did but setting it up is too complicated) try the IM+ download. It supports all the major IM services including AOL, ICQ, MSN, and Yahoo!. Here's how to set up IM+:

1. **Point your PC browser to SymbianWare** (www.symbianware.com).

2. **Download and install the IM+ instant messaging client using the steps in Technique 72.**

3. **On your phone's main menu, select IM+.**

4. **Inside IM+, press Menu on the left menu key.**

5. **Select Options.**

6. **Select your service from the list.**

7. **Enter the username and password that you use to sign in to the service.**

   All the other settings (server, port, attempts to reconnect, and so on) are preset.

8. **Press Back on the right menu key to return to the main screen.**

9. To connect, press Menu on the left menu key.

10. Select your service from the list.

11. Select Available/Connect (see Figure 12-22).

Figure 12-22: Select Available/Connect to go online with your IM service.

The first time you connect, you'll have to choose which access point to use, but you'll be shown a list from the access points configured on your phone (using Technique 2), so just pick one.

Your existing buddy list will automatically be downloaded to your phone. Just select a buddy and start chatting. It's that simple. As always, try IM+ before you decide to buy it. See if it works properly for you and your IM service.

Exploiting Useful Tools

Your Series 60–based phone is a powerful tool straight out of its box. But you can make it even more powerful by adding software tools and utilities to it. Whether you work in business or academics, or you just use your phone as a communications portal, you'll find something of interest in this chapter.

Here I describe a package that lets you view on your phone documents created with Microsoft Office, how to compress to and extract from WinZip files, how to control your phone using your PC, and more.

Also, a special note for people with visual impairments — from slight to significant. I describe a great text-to-speech tool specifically designed for Series 60 smartphones that lets you use your phone without ever needing to look at it.

♦ ♦ ♦ ♦

In This Chapter

Viewing Word, Excel, and PowerPoint files on your smartphone

Using compression software to store and transfer files without using much memory

Discovering how software can empower the visually impaired or anyone needing to use their phone "eyes free"

♦ ♦ ♦ ♦

Technique 81: Viewing Word, Excel, and PowerPoint Files on Your Smartphone

The popularity of Microsoft Office applications — like Word, Excel, and PowerPoint — is astounding. People use these formats for all types of business and academic pursuits, even for running a home. In Technique 33, I show you how to transport documents like these on your phone to and from work and home.

But wouldn't it be nice if you could also view them while they were on your phone? You can! Now, not only can you carry your work with you on your smartphone, you can do work.

The tool is called QuickOffice, and it's a set of three separate viewers for Word documents, Excel spreadsheets, and PowerPoint presentations.

Tip

QuickOffice requires more memory than most Series 60 applications. Before I could install and run QuickOffice, I had to perform some of the memory cleanup techniques described in Techniques 64 and 97. Save yourself the time involved in a failed installation (too little memory to complete the task) by doing a backup and a cleanup before you start the QuickOffice installation.

If you do encounter a low-memory warning during the installation, cancel the installation by pressing Cancel on the right menu key. Then, use your cleanup techniques to free some space in phone memory or your memory card (if you're planning to install it there). And, finally, restart the installation process.

Note

QuickOffice, at about $50, costs a little more than most of the applications mentioned in this book. So, trying it before you buy it is even more important. Download the free trial version and experiment with it on your phone with your documents.

Download QuickOffice to your PC from `www.quickoffice.com`. Then perform Technique 72 to install the application to your phone using your phone's PC software.

Note

An alternative installation method: Send QuickOffice over Bluetooth, IR, or USB through a direct connection to the phone and install from the "Messaging" inbox.

The QuickOffice installation places the three file viewers on your main menu as three separate applications (see Figure 13-1). Unless you use them very frequently, I recommend moving them into your Tools folder or another folder of your choosing using Technique 37.

Figure 13-1: QuickOffice's Word, Excel, and PowerPoint viewers.

What is a viewer?

A *viewer* is a software application that allows you to look at — but not edit — a file of a particular format.

Microsoft Word, Excel, and PowerPoint each use a particular format that cannot be opened with a simple text editor. Also, currently, none of these three application types is supported by Series 60 phones, so for a while now, Series 60 smartphone users were seemingly stuck. But no longer.

Enter QuickOffice. QuickOffice supplies "viewer" applications for Word, Excel, and Powerpoint for the Series 60 platform — filling this previously unmet need.

Technique 82: Creating and Using Compressed WinZip Files on Your Phone Using the ZipMan Download

File compression technology remains an important part of computing. Whenever you send large files over e-mail or transfer files from one computer to another, it's a good idea to compress them to a smaller size so that the sending or transfer "pipe" doesn't get clogged with data and slow things way down.

There are several compression tools and formats, but one of the most popular is called WinZip. WinZip lets you create compressed archives of files. You can then attach these archives to e-mails or send them over a network. Because the compressed files are smaller than the combined sizes of the original files, moving them from one computer to another is easier and faster.

Tip Use ZipMan in conjunction with Technique 33 about using your phone as a temporary storage device. Now you can store more in less space using compression.

You can use ZipMan in two basic ways:

✦ You can create compressed ZIP files using a PC tool like WinZip, and then you can transfer it to your phone and open it with ZipMan.

✦ You can create new compressed archive files right on your phone using ZipMan to transfer them to a PC or another device.

Here's how to open a ZIP archive file that you can send to your phone:

1. **Create an archive file on your PC using WinZip (www.winzip.com).**

2. **Transfer the file to your smartphone using Technique 52.**

   I recommend putting the archive right into the ZipMan folder on your phone.

3. **On your phone's main menu, select ZipMan (see Figure 13-2).**

Figure 13-2: Select ZipMan on your main menu.

ZipMan finds any ZIP files resident on your phone and displays them in its archive list (see Figure 13-3).

Figure 13-3: ZipMan's archive list.

4. **Select the archive you want to open.**

   Now you'll see a list of the files inside the archive. To save an individual file or all the files within the archive to your phone you need to extract it or them.

5. **Press Options on the left menu key.**

6. **Choose Extract (see Figure 13-4).**

Figure 13-4: Choose Extract to save a file from the archive to your phone.

7. **Choose All Files to extract all the files within the archive (see Figure 13-5).**

Figure 13-5: Choose All Files to extract all the files to your phone.

8. **Navigate to the folder where you want to extract or save the files.**

9. **Press OK.**

This will save all the files directly to your phone memory or memory card. You can use them as you normally would.

Tip

You can purchase extensions to ZipMan that let you open GZip, Tar, and Rar compressed files (popular on Unix/Linux environments). Browse WildPalm (www.wildpalm.co.uk) for more information.

To create a new ZIP file on your phone, do the following:

1. **On the main menu, select ZipMan.**
2. **Press Options on the left menu key.**
3. **Choose New Archive (see Figure 13-6).**

Figure 13-6: Choose New Archive to create an archive on your phone.

You'll see a blank list for this new archive (its name will be automatically set to ZipArchive.zip, but you can rename it using Rename on the Options menu).

4. **To add files to the archive, press Options.**
5. **Choose Add File.**
6. **Navigate to a file and select it.**
7. **Select OK when you're done.**

Note

You can set a password on an archive file and send the archive to a PC or another device using the Options menu within ZipMan.

With compression, you can stretch the memory on your phone and speed up the sending and receiving of large files. You can download ZipMan from WildPalm (www.wildpalm.co.uk). As always, though, try it before you buy it!

What is compression technology and how does it work?

Compression technology works by algorithmically removing repeated information from a file. Certain types of files have more repeated information than others. For example, image files (think of a blue sky — at a pixel level this is just the same information over and over again — blue, blue, blue, blue) generally have more repeated information than word-processor documents. And, word-processor documents generally have more repeated information than applications.

When you compress files, you'll see compression percentages similar to the following ranges:

✦ Applications get compressed by approximately 10 to 30 percent.

✦ Word-processor documents get compressed by approximately 40 to 70 percent.

✦ Image files get compressed 60 to 90 percent.

The compression percentages are not critical information for you to know. But it *is* helpful for you to know that a folder full of image files will become much smaller when you compress them, while a folder full of applications likely won't become all that much smaller.

Technique 83: Taking Pictures of Your Phone's Screen Using the ScreenTaker Download

Pictures are worth approximately 1K words — in techspeak. That's why this book is filled with images of a phone's screen. I show you as well as tell you what's going on.

I'm sure there will be times when you yourself will want to show someone what's going on with your phone's screen. Maybe you want to print out some contact information, capture a great moment in game playing — or maybe you have a smartphone tip of your own that you'd like to share.

If so, I'll let you in on a little secret: I captured all the phone screen images in this book with a tool called ScreenTaker. You can download it from SymbianWare (www.symbianware.com) and install it on your smartphone using Technique 72. It's one of the few screen-capture utilities currently made for Series 60 platform–based handsets.

After you install it, here's how you use ScreenTaker:

1. **On the main menu, select ScreenTaker (see Figure 13-7).**

Figure 13-7: Select ScreenTaker to set up your phone screen capture.

2. **Press Hide on the left menu key to put ScreenTaker into the background.**

   It will remain active even though it won't be visible.

3. **Bring up the screen you want to capture.**

   It can be anything—a contact record, a pop-up window, a game's screen, whatever.

4. **While holding down the Edit key (it looks like a pencil on most Series 60 phones from Nokia, the up-arrow on Siemens SX1, or a button on the left side of Sendo X) on your phone, press the * key.**

   Yes, this may require two hands.

   This invokes the screen capture and brings up ScreenTaker's Image Name pop-up. By default, the name of each image is SCR.jpg (the format is a compressed JPG).

5. **You can change the name to anything you like.**

   It's probably best to keep the JPG file type in the name because many applications both on your phone and on a PC use that file type to decide how to open the file.

6. **Press OK when you've finished editing the name.**

   ScreenTaker shows a Image Saved! pop-up.

7. **Press Back on the right menu key.**

8. **To end screen capturing, select the ScreenTaker object again in the main menu and press Exit on the right menu key.**

Technique 84: Controlling Your Phone via Your PC Using the Remote S60 Pro Download

The Series 60 platform's user interface is very user-friendly, but it's hard to beat a full-size desktop computer keyboard and mouse for comfortable data entry and navigation through menus.

Guess what? With a software download called Remote S60 Pro, you can drive your phone using your PC's keyboard and mouse. You can even see the phone's screen on your PC's screen, so you could actually leave your phone resting snugly in your backpack or belt case.

Download Remote S60 from SymbianWare (www.symbianware.com) and yes—say it with me now—try before you buy!

Remote S60 comes with a PC application and a phone application. The two talk to each other using whatever connection you specify. It can connect the phone using infrared or Bluetooth. A Bluetooth connection is faster and offers better range than infrared (you can keep the phone in your pocket with Bluetooth).

After you've installed both the PC and phone software, following the installation instructions, do the following to start using the software:

1. **On the phone main menu, select Remote S60 Pro.**

2. **Press Options on the left menu key.**

3. **Choose Connect.**

4. **Choose Bluetooth (in this example).**

   On your PC, you'll see a request from a device for a Bluetooth serial connection.

5. **Click OK.**

6. **Launch the Remote S60 Pro application on your PC.**

   Now, on your PC's screen you'll see your phone's screen (see Figure 13-8).

7. **I recommend bringing up a "skin" on the PC application to make the window look like a phone. Do this by choosing View ➪ Skin ➪ Nokia 6600.**

   Now, you'll see your phone's screen on your PC, but it will be wrapped by a Nokia 6600 skin (see Figure 13-9).

Figure 13-8: Your phone screen on your PC through Remote S60 Pro

Figure 13-9: The transformation is complete. Your phone is up and running on your PC screen.

At this point, it's all up to you. Experiment with your keyboard and mouse. You're in complete control. You can browse and edit your contacts and calendar, send MMS and SMS messages, and more.

Technique 85: Making Your Phone Talk to You

I've always believed in computer technology's ability to empower those with disabilities. If you've ever seen someone who has a disability do something he's never done before with the help of a computer, you'll believe it, too.

Now that phones have begun to evolve into smartphones, you can download and run software, and that's all you need to do to make your phone talk to you. For those with visual impairments, this can sometimes mean the difference between using a mobile phone and not.

The application in this technique is called TALKS. It's an application developed for Series 60–based smartphones that uses a ScanSoft text-to-speech (TTS) engine. To get your own copy, point your PC browser to www.talx.de/index_e.shtml. Click on the Dealers/Links link to find a distributor in your country.

Tip   This category of technology is called *accessibility technology.* For more information on mobile accessibility technology, point your browser to Nokia's Accessibility Web site (www.nokiaaccessibility.com).

When you purchase the software, you'll need to provide your serial (IMEI) number so a software license can be generated that is specific to your smartphone. Read Technique 98 about a secret code that will show you your phone's serial number (IMEI number) on the phone's screen.

Follow the application instructions for downloading. Install it on your phone using the steps from Technique 72. The installation will install both ScanSoft's TTS engine and the TALKS application. The installation routine installs three objects on your main menu: Talks, Talks On, and Talks Off (see Figure 13-10).

Figure 13-10: The installation puts Talks, Talks On, and Talks Off objects on your phone's main menu.

Here's how to set up TALKS:

1. **On your main menu, select Talks.**

   It contains the settings for the application.

   Note As soon as you select Talks, you'll hear your phone begin talking to you. This application reads most of the things on your screen out loud to you so that you can use your phone without looking at it at all. It announces all callers out loud by reading the caller ID information to you. It reads your SMS messages to you. It will tell you where you are in your contacts list or calendar, even as you're scrolling through. And it will always tell you what the left and the right menu keys do.

2. **Read the documentation for detailed information about all the settings.**

   I won't go into them in detail here. You'll probably find you can use TALKS with its default settings. But I want to show you three of the settings that you may want to adjust so that you can hear the phone's talking more clearly:

   - **Volume:** Select this and press your joystick to the right or left to raise or lower the volume until it's at a comfortable setting for you.

   - **Speed:** Select this and press your joystick to the right or left to make the speaker talk faster or slower. Make the speed comfortable for you.

   - **Pitch:** As with the other two settings, adjust the pitch of the voice up or down until it sounds as nice as a close friend.

3. **Press Exit on the right menu key when you've finished adjusting these and any other settings.**

To start using TALKS couldn't be easier. Simply select Talks On on your phone's main menu. To stop Talks from speaking simply select Talks Off.

Anyone who needs to use his phone "eyes-free" or without looking at it can make use of TALKS. I use it fairly frequently while driving in conjunction with a Bluetooth wireless headset. I never have to take my eyes off the road because I can just listen to what SMS message has come in or what my upcoming appointments might be. I promise you'll be amazed at what this text-to-speech application can do.

Managing Your Time

PDA, schmee-D-A. You may think you need a Palm, PocketPC, or other personal digital assistant (PDA) to help you stay on top of your meetings, appointments, and anniversaries — but you'd be wrong.

Blame your ignorance of your smartphone's calendar on its user manual or the fact that its cool features aren't covered in high-tech magazines and Web sites. Then read through this chapter to learn the ins and outs of Series 60–based time management. By the end of this chapter, you'll know how to synchronize your PC and phone calendars, get reminders from the Web, customize your calendar views, and more.

And you won't need a bigger briefcase, backpack, or purse to carry bulkier devices — your smartphone stores everything you need to stay on schedule.

Technique 86: Receiving Calendar Reminders from the Yahoo! Calendar

Do you use a Web-based calendar like Yahoo! Calendar (see Figure 14-1)? It's a very popular, free service that lets you post appointments and access your schedule from any PC that can access the Web. Even better, Yahoo! Calendar (http://calendar.yahoo.com) allows you to send meeting and appointment reminders to your smartphone by SMS.

In This Chapter

Synchronizing your PC's calendar with your phone's calendar

Changing your calendar views

Setting and using calendar alarms

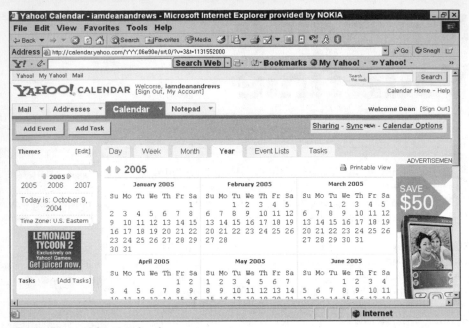

Figure 14-1: Yahoo! Calendar

For usability, I suggest setting your Yahoo! Calendar appointments and reminders via a PC, not over your phone. But, hey, if you feel like spending a few extra minutes putting in that anniversary reminder via your phone's Web browser, go for it!

In the following example, I show you how to schedule an anniversary reminder for August. Here are the steps to schedule an appointment and register your smartphone with Yahoo! Calendar:

1. **Point your PC Web browser to** `http://calendar.yahoo.com`.

2. **Use the Day, Week, Month, and Year tabs, and the left and right arrows, to navigate to the date of your new appointment.**

   In this example, I'm using August 29.

3. **Choose [Add] under the appropriate day (see Figure 14-2).**

   The Add Event screen appears.

Figure 14-2: Add a new appointment in Yahoo! Calendar.

4. **Enter the Title, Event Type, Time, Location, and Notes for your appointment.**

5. **Scroll down the page to the Reminders section.**

6. **Select the check box for Mobile Device Email, and then click the Add Mobile Device link.**

7. **Select Mobile Phone in the Device Type drop-down list.**

8. **Make a choice in the Daily Message Limit drop-down list (see Figure 14-3).**

   If you have a limit on the number of text messages you can receive in a day or month — based on your service plan — choose a lower number here.

9. **Click Continue to Step 2.**

10. **Enter your ten-digit phone number, and select your service provider and phone maker from the drop-down lists (see Figure 14-4).**

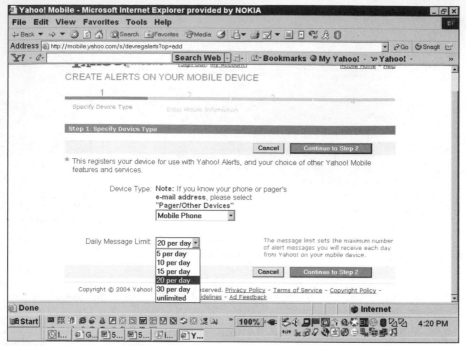

Figure 14-3: Choose the right message limit to suit your cellphone service plan.

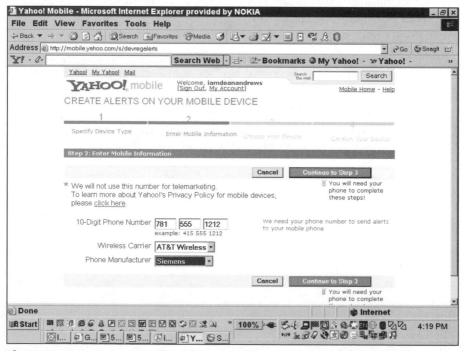

Figure 14-4: Enter your ten-digit phone number, service provider, and phone maker.

11. **Click Continue to Step 3.**

12. **Choose your phone from the set of pictures (see Figure 14-5).**

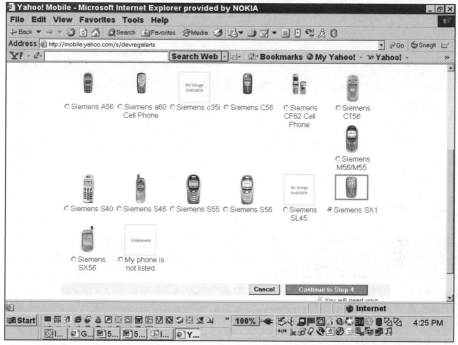

Figure 14-5: Choose your phone from the set of pictures.

13. **Click Continue to Step 4.**

   At this point, Yahoo! Calendar sends a confirmation code in a text message to your phone.

14. **View the text message on your phone and enter the code in the Step 4: Confirm Your Device page of the add device process (see Figure 14-6).**

15. **Click Confirm.**

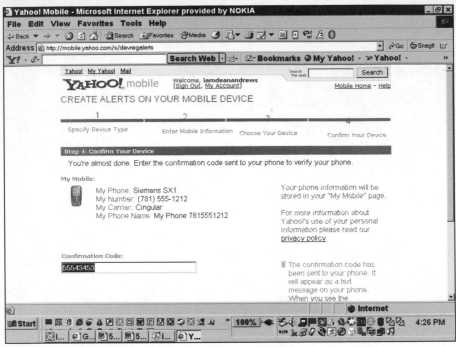

Figure 14-6: Enter the confirmation code that Yahoo! Calendar sends in a text message to your phone.

Now whenever you schedule a new appointment in Yahoo! Calendar, you'll be reminded via a text message directly to your phone.

Tip Use the options under Reminders to set how far before an appointment the text message is sent.

Technique 87: Using Calendar Alarms

Having your calendar reminders sent in from the Web or through synchronization with your PC is great, but sometimes you may need to enter new appointments right on your phone. Fortunately, doing so is quick and easy because of the well-designed calendar interface on your smartphone.

On your smartphone, a calendar reminder is not only a description of an appointment but also an alarm with audio and visual effects to get your attention — all the better to keep you on schedule in your busy day.

Jumping to a particular day

Is the day you want more than a month away from your current position in the calendar? You may want to jump directly there rather than scroll a day or week at a time using the joystick. How? Follow these steps:

1. Inside Calendar, choose Options ⇨ Go To Date (see the figure).

Go to Date takes you directly to a specific date.

2. Enter the date to which you want to go (see the figure).

Pressing the joystick moves the cursor to the different fields: Month, Day, Year. Pressing the joystick up and down moves the month, day, or year forward or backward, depending on where the cursor is positioned.

Enter your date.

3. Press OK on the left menu key to jump to that date on the calendar.

You can select the day at your new location in the calendar in order to add an appointment or a memo.

When the cursor is on a day in your calendar, just start typing. The calendar automatically displays a meeting reminder and then puts the text you're typing into the subject field.

Here's how to create a new calendar reminder:

1. **On the main menu, select Calendar.**

   The default Calendar view is a whole month at a time (see Figure 14-7). In Technique 88, I show you how to customize this view to your liking.

   Figure 14-7: The calendar's month-at-a-time view

2. **Using the joystick scroll to the date in which you want to add an appointment.**

   Note In month view, pressing the joystick left or right moves the cursor one day forward or backward. If the next (or previous) day is in a different month, the screen rolls to the appropriate month. Pressing the joystick up or down moves the cursor back or forth one week (on the same day of the week from which you start). If the next (or previous) week is in a different month, the screen rolls to the appropriate month.

3. **Select the day for which you want to add an appointment.**

   You could also choose Options ⇨ Open to view a particular day — but why waste the key clicks?

4. **Press Options on the left menu key.**

5. **Choose New Entry (see Figure 14-8).**

6. **Choose the type of reminder you want: Meeting, Memo, or Anniversary (see Figure 14-9).**

7. **Enter the information in the fields of the reminder.**

8. **Press Done on the right menu key when you're finished entering a new appointment's details.**

Figure 14-8: Choose New Entry.

Figure 14-9: Choose the
type of meeting you want.

 Tip

To send an appointment to someone else, highlight a particular appointment (see Figure 14-10), and then choose Options ⇨ Send. You'll be asked whether you want to send the appointment via a text message, via a multimedia message, or through a Bluetooth connection to another phone or device (see Figure 14-11).

Figure 14-10: Highlight a
particular appointment.

Figure 14-11: Choose Options ⇨ Send and you'll be presented with a list of sending methods.

Setting a calendar alarm

While entering a new calendar reminder, you'll encounter the Alarm field. Alarm, by default, is set to Off. Here's how to set it:

1. Select Alarm.

2. Choose On (see the figure).

Choose On to set a new alarm.

Now, two additional fields — Alarm Time and Alarm Date — will appear.

3. Scroll down to Alarm Time.

By default the Alarm Time is set to 15 minutes before the appointment's start time (see the figure).

4. Use your joystick to adjust the hour and minutes (and AM or PM).

5. Scroll down to Alarm Date.

The Alarm Date is automatically set to the day of your appointment.

6. Use your joystick to adjust the date to whatever you want.

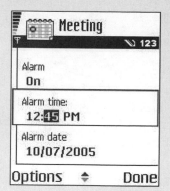

The alarm time is automatically set to 15 minutes before the start time of your appointment, but you can change it whatever you like.

That's it for setting an alarm. If you want to change the tone used for alarms, choose Options ⇨ Settings ⇨ Calendar Alarm Tone. You can then pick from a list of tones and tunes available on your phone.

When your alarm sounds, press any key to put the alarm into five-minute snooze mode. Or press Stop on the left softkey to permanently silence the alarm.

Technique 88: Customizing Your Calendar Views

You know how traditional calendars come in different formats and styles? You can buy day-by-day or week-at-a-glance calendars for your desk, or you can purchase regular month-by-month calendars that you hang on the wall. Did you know you have the same style choices for your smartphone's calendar?

For a quick look at Day, Week, and Month calendar views, see Figure 14-12, Figure 14-13, and Figure 14-14, respectively.

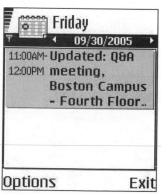

Figure 14-12: Day view

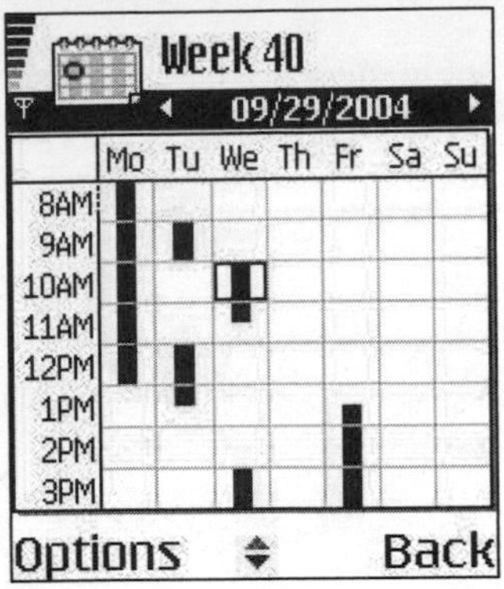

Figure 14-13: Week view

Figure 14-14: Month view

Follow these steps to set the default view for your phone's calendar:

1. **On the main menu, select Calendar.**

2. **Press Options on the left menu key.**

3. **Choose Settings (see Figure 14-15).**

4. **Select Default View (see Figure 14-16).**

> **Tip** Jump to today's date in your calendar by pressing the # key.

5. **Choose the view that best suits your lifestyle (see Figure 14-17).**

Figure 14-15: Choose Settings.

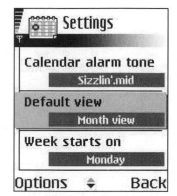

Figure 14-16: Select
Default View.

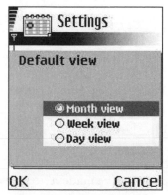

Figure 14-17: Choose from
Month, Week, or Day views.

Tip

You can switch views on the fly too. Say your default view is set to Month, but you want to quickly look at just the week ahead of you. Inside your Calendar, choose Options ➪ Week View (see Figure 14-18). You can also choose Options ➪ Month View to get the "big picture" whenever you want.

Figure 14-18: Leave your default view set to Month, but quickly scope out your week by selecting Options Í Week View.

6. **Press OK on the left menu key.**

But, wait! There's more! Are you picky about which day of the week your calendar starts? I know that drives me crazy. I sometimes pencil appointments into the wrong day if the calendar I'm working with starts on a Saturday or Sunday rather than a Monday.

Well, your smartphone calendar doesn't have to drive you crazy no matter what day it shows first. You can set it to any day you like:

1. **On the main menu, select Calendar.**

2. **Press Options on the left menu key.**

3. **Choose Settings.**

4. **Select Week Starts On (see Figure 14-19).**

5. **Pick your favorite start day (see Figure 14-20).**

Looks pretty good, doesn't it? That's because now you've made it your own.

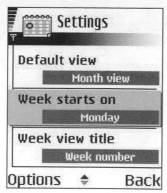

Figure 14-19: Select Week Starts On to reformat your calendar.

Figure 14-20: Set your calendar's start day.

Technique 89: Synchronizing Your Phone and PC Calendars

Who wants to keep more than one calendar up-to-date? I can barely manage keeping one calendar current. Fortunately, you don't need to juggle more than one schedule. In fact, you can make updates either in your phone's calendar or in your PC's calendar — whichever is most convenient. Then use the power of synchronization to share the most current schedule between your phone and PC. Here's how:

1. **On your PC, launch your phone's PC software (see Chapter 9 for more information on using phone PC software), as shown in Figure 14-21.**

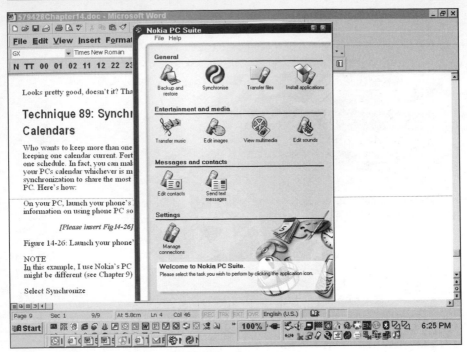

Figure 14-21: Launch your phone's PC software.

 Note   In this example, I use Nokia's PC Suite Version 6.0. Your phone's software may be different (see Chapter 9), but it will work in a similar way.

2. **Select Synchronize (see Figure 14-22).**

   The first time you use the synchronization software, you'll need to configure it. Read Chapter 9 for instructions on configuring your PC software to synchronize with a PC application.

3. **Press the Synchronize Now button.**

   You'll probably have to accept a connection request on your phone. A progress screen will show what is happening moment by moment.

   The synchronization will happen in most cases without your having to do anything. You'll see a confirmation alerting you when the synchronization is complete (see Figure 14-23).

I try to synchronize my phone and PC calendars once a day, but I frequently add appointments both on my phone and on my PC. If you have less-frequent changes, you may only need to synchronize once a week.

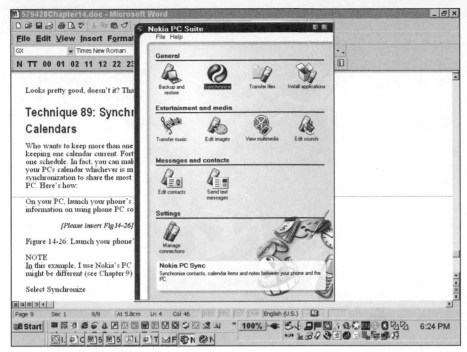

Figure 14-22: Select Synchronize.

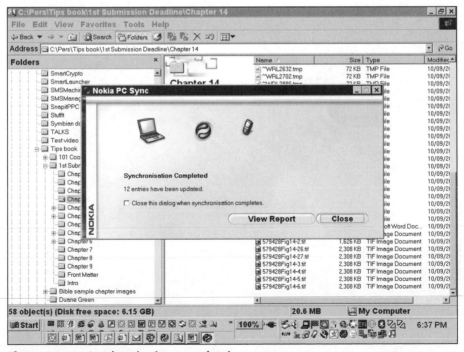

Figure 14-23: Synchronization complete!

Browsing the Web with Your Phone

Browsing the Web over your phone is a little like
watching the NFL's world championship, the
Super Bowl. Before it begins, you feel hyped up, and the
event seems filled with possibility. Every once in a
while, the event lives up to these expectations. But,
more often than not, the event itself ends up being
tedious, boring, and anticlimactic.

Why is Web browsing over your phone hit or miss? I
think the problem still revolves around usability. When
compared to Web browsing over a PC, phone browsing
means a much smaller screen, potential lack of a full
QWERTY keyboard (or in some cases a very small one),
no mouse, and in most cases, slower access speeds.

How can you make phone browsing better? First, learn
how to configure and use your phone's browser properly. Your Series 60–based phone also allows you to
download a new browser. Then, learn how to pick and
choose the sites you browse over your phone and save
them as bookmarks. Finally, explore the latest methods
for easy and fast access to Web information via
phone — like a very cool new idea called Semacode.
This chapter shows you how to do all these things.

Technique 90: Configuring and Using Web Browsers

If you haven't already done so, read Technique 2 about
how to request a special configuration SMS to automatically configure your phone's Web browser for Web
access through your service provider. It's fast and
easy to do.

Tip   While your Nokia phone is in standby mode, press and hold the 0 key to automatically launch your Web browser. This shortcut doesn't work on non-Nokia Series 60–based phones.

Tip   Ever try sending a bookmark from your phone to someone else? Save a friend the hassle of typing in a URL over his phone's keypad by sending your favorite sites to him in a text message. While highlighting a bookmark on your browser's bookmark screen, choose Options ⇨ Send (see Figure 15-1). Then, select Via Text Message. The bookmark's URL will be placed inside a message in the Message Editor. Add a Recipient in the To field and send the message as you normally would.

Figure 15-1: Choose Options ⇨ Send to send a browser bookmark from your phone to another phone.

Note   The browser application is titled differently on different phones. It's called Internet on the Siemens SX1; WAP on the Sendo X; Services on the Nokia 7650, 3600/3620/3650/3660, 6600, N-Gage, and N-Gage QD; and Web on the Nokia 6260, 6620, 6630, and 7610 and the Samsung SGH-D710.

After your browser can access the Web, you'll want to adjust its settings to suit you. Here's how:

1. On the main menu, select Web.

2. Inside the browser, press Options on the left menu key.

3. Choose Settings (see Figure 15-2).

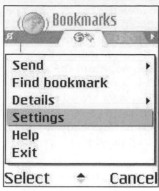

Figure 15-2: Choose
Settings.

Now you'll see the browser's list of settings. I won't mention every
one, just the ones you should know:

- **Default Access Point:** This should have been set by the configuration message from your phone maker, so you won't need to
change it.

- **Show Images:** If you're dissatisfied with the speed that Web
pages display on your phone, change this setting to No. You can
always choose to load the images of a Web page while you're
viewing it by choosing Options ➪ Show Images. If you aren't dissatisfied with the speed of page loading, keep this setting at Yes.

- **Text Wrapping:** Setting this to On ensures you'll see all the text
of a link even if the link wraps around to more than one line of
the display.

- **Font Size:** Changing this to Small or Smallest is a little like
increasing the resolution on your PC. In theory, more stuff will fit
on the screen. In some cases, you won't notice the difference.
Text embedded into graphics, for example, won't be changed by
this setting.

The following features are not supported in all Series 60–based phones.
Only the Nokia 6260, 6620, 6630, and 7610 and the Samsung SGH-D710
support the following three features.

- **Automatic Bookmarks:** This works like Internet Explorer's auto-
complete feature. It helps you with the hassle of entering a new
URL by trying to complete the URL for you by "guessing" what
words you're typing. I recommend you set this On.

- **Screen Size:** This setting adjusts the area the current Web page consumes on your screen. Normal screen (see Figure 15-3) shows the Web page in the middle, your browser title bar at the top, and the selection keys on the bottom. Selection Keys Only screen (see Figure 15-4) shows the Web page and the selection keys at the bottom but no title bar. Full Screen (see Figure 15-5) shows just the Web page. The selection keys are still active, but they aren't displayed. If you're new to Web browsing by phone, I recommend making this setting Selection Keys Only. If you're experienced at phone browsing, I recommend Full Screen.

- **Rendering:** By Quality means you'll have higher-quality images but slower downloading. By Speed means faster downloads with slightly lower-quality images. I recommend By Speed.

Figure 15-3: The Normal screen size

Figure 15-4: The Selection Keys Only screen size

Figure 15-5: The Full
Screen size

After your browser is configured and the settings adjusted to your liking, you'll
want to start browsing. My advice is basically this: Get URLs sent to your
phone, enter a few that you like, and then save them all as bookmarks. From
then on, your phone browsing will be a matter of clicking on bookmarks with
your joystick, thereby removing most of the hassle from the experience.
Here's how:

1. **Use the tip at the beginning of this technique to have your friends
   and colleagues send their favorite Web bookmarks to your Series
   60 phone.**

   You can also request bookmarks from your phone's maker and your
   service provider. Typically, these companies create and update their
   own favorite bookmarks and don't let you choose which you want
   sent to your phone, but others do let you choose. Access your
   phone-maker and service-provider Web sites for details.

2. **On your main menu, go to Messages.**

3. **Select Inbox.**

4. **Highlight the message that contains a URL and open the message
   (see Figure 15-6).**

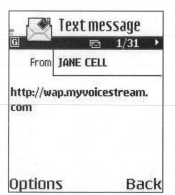

Figure 15-6: Open the message
that contains a URL.

5. **Press Options on the left menu key.**

6. **Choose Find (see Figure 15-7).**

Figure 15-7: Choose Find.

7. **Choose Web Address (see Figure 15-8).**

Figure 15-8: Choose
Web Address.

The URL will appear highlighted in the message (see Figure 15-9).

Figure 15-9: The URL appears highlighted in the message.

8. **Press Options again.**

9. **Choose Add to Bookmarks (see Figure 15-10).**

Figure 15-10: Choose Add to Bookmarks.

You'll receive a confirmation message that the bookmark was saved. The default name is "bookmark."

10. **Launch your browser, highlight the desired bookmark, choose Options ⇨ Edit, and change the name to a memorable one.**

Note that the Nokia 6260, 6620, 6630, and 7610 and the Samsung SGH-D710 have a Bookmark Manager feature that results in a slightly different navigation sequence from the preceding. On these phones, it's Options ⇨ Bookmark Manager ⇨ Edit.

11. **Inside your browser's bookmark page, simply select a bookmark to go to its link.**

Want to try another Web browser on your phone? The Opera browser is a popular alternate choice.

Note   The Series 60 Web browser is tightly integrated with the Series 60 user interface. Other applications — like Clock, which can be set to get time updates from the Web — use the browser under the cover to synchronize with the Web. Another example of this integration is the ability to add Bookmark shortcuts to the Favorites/Go To application from the browser. When you install another browser, it won't be as tightly integrated as your standard Series 60 browser.

You can download the Opera browser from www.opera.com. Make sure you pick the correct version of the browser for your phone. Install it on your phone using Technique 72. The Opera browser features a different user interface and different keyboard shortcuts. Opera also offers a search engine and news portal that you access from the phone browser. Preferring one phone browser to another is ultimately a matter of personal choice.

Tip   Have you heard about Moblogs (Mobile Web Logs)? Moblogs are basically Web sites that allow you to upload images (and other multimedia) from your mobile phone, annotate the pictures with text, and share them with others over the Web. A Moblog is like a high-tech multimedia diary with an audience of millions.

To get started, check out these popular Moblogs: Buzznet (www.buzznet.com), Mobog (www.mobog.com), and TextAmerica (www.textamerica.com).

Also, check into Nokia's cool LifeBlog (www.nokia.com/lifeblog), which lets you create a Moblog across your smartphone, PC, and the Web.

Technique 91: Accessing a Web Link by Snapping a Digital Picture

Let's hear a round of applause for application developers! These intrepid engineers are always looking ahead and inventing the future.

Remember what I said in the last technique about the difficulty of browsing via phone because of the problem of entering text (or URLs)? Well, check out this cool new method of accessing Web sites via the phone. It's called Semacode (www.semacode.org). It takes advantage of your smartphone's camera and its ability to access the Web, and it addresses the problem of entering long URLs on a small phone keypad.

Tip   For another clever way to access Web information over your mobile phone, read up about Google SMS (www.google.com/sms/). Basically, the popular search engine Google allows you to search the Web and get results all through text messaging through this service.

Here's the basic concept: With your camera, you can snap a picture of a special coded square—a Semacode—and, automatically, your phone's browser will display the Web site to which the coded square told it to go.

These Semacodes can be printed on paper, put onto stickers, or printed in books and magazines. You can also create your own Semacodes.

Say you're looking through a magazine you just bought off a newsstand. While reading a movie review, you see a little square in the corner of the article next to a sentence that says, "If you want to see the movie's trailer, snap a picture of the Semacode with your smartphone's camera."

You don't have to wait for this new method to gain popularity. You can use it right now:

1. **Point your PC browser to** `www.semacode.org`.

2. **From the site, download the Semacode reader for your phone and install it on your phone using Technique 72.**

3. **From the site, download the Opera browser for your phone and install it on your phone using Technique 72.**

    The Semacode reader was designed to work with the Opera browser.

4. **On your phone's main menu, select Semacode (see Figure 15-11).**

Figure 15-11: Select the Semacode reader.

5. **Using the sample Semacode square found at** `www.semacode.org` **(either print out the Semacode on paper or display the image on a flat LCD screen), snap a picture of the sample Semacode using the Semacode reader (press the joystick down to capture and read the Semacode) (see Figure 15-12).**

Figure 15-12: Capture and read the Semacode.

You'll see a decoding message. Then, you'll see the URL the reader decoded from the Semacode (see Figure 15-13).

Figure 15-13: The reader decodes the URL.

6. **Press OK.**

The Opera browser launches automatically and takes you to the URL.

Make your own Semacode images using the Semacode Creator at the Semacode Web site. Then stick them on posters, print them on envelopes, or whatever. Spread the word about this cool new way to browse the Web with a smartphone!

Gaming

Games have been available on cellphones from almost the very beginning. I have fond memories of playing Snake on an old Nokia phone during my daily commuter-train trips. The game had only the most basic of graphics, no color, and simple one-note-at-a-time sound effects — but it was still fun. And very addictive.

Along with everything else on your phone, though, phone-based games have evolved. Now phone games feature complex graphics, full color, full sound tracks — and they can be downloaded straight from the Web.

In this chapter, I describe how to download games directly from the Web to your phone, clue you into the latest trend in phone gaming, and provide a list of some of the best sources for smartphone games.

Technique 92: Downloading Games

Quick, what's the hottest trend in phone gaming? If you said, "Games that make use of your phone's integrated hardware," step to the head of the class.

There are a lot of choices in mobile game platforms, but some of the key elements that differentiate your Series 60–based smartphone from the usual suspects are the Bluetooth personal network capability, the integrated digital camera, and the microphone. The most clever game developers have realized this, and they've begun producing games that take advantage of this hardware difference.

Games like Bluetooth Darts (see Figure 16-1) allow for multiplayer gaming across phones using Bluetooth wireless networking. And, Mozzies (available on the Siemens SX1) uses your phone's digital camera as a gun's crosshairs.

Figure 16-1: Bluetooth Darts allows two players on two different smartphones to play together.

If you're tired of the same games, look for these new creative variations on the themes that were specifically designed for your Series 60–based smartphone. These types of games aren't hard to spot. When you browse the sources listed in Technique 93, read the short descriptions that explain the game's highlights. You'll see mention of Bluetooth, integrated camera, and microphone there, if the game takes advantage of them.

If you've read Techniques 72 and 73, you know how to download Symbian and Java games to your PC and install them on your phone using your phone's PC software. Because your phone can access the Web, however, you can download cool games directly from the Web to your phone. Here's how:

1. **On the main menu, select Web (or Services).**

2. **Point your browser to a gaming Web site.**

   In this example, I'm going to www.funkyplaza.com. FunkyPlaza is a demo site, available to the world, that was established by Nokia to show individuals some of the multimedia content available for smartphones to access and download. It features games, video, music, and other content.

3. **Scroll down to the Game Planet link (see Figure 16-2).**

Figure 16-2: Scroll to Game Planet.

4. **Select Game Planet.**

5. **Select the Hot Downloads link (see Figure 16-3).**

Figure 16-3: Select
Hot Downloads.

6. **Select Yo Yo Fighter (see Figure 16-4).**

Figure 16-4: Select
Yo Yo Fighter.

7. **Select Download (see Figure 16-5).**

Figure 16-5: Select Download.

You'll see the download progress bar (some phones don't show the progress bar) at the top of the browser along with an increasing count of kilobytes.

8. **From here, follow the instructions on the installation pop-up screens to complete the installation.**

After its installed, you can find the downloaded game at the bottom of your main menu.

Technique 93: Searching for Smartphone Games on the Web

As with technical support, your phone maker and service provider Web sites should be your first stops for new phone games. These two companies have a vested interest in making your phone experience as good as it can possibly be. Almost all phone makers and service providers now offer games for download.

But there are other great sources out there on the Web. Here are some good examples:

✦ **Software Market:** www.softwaremarket.com (see Figure 16-6)

✦ **Handango:** www.handango.com (see Figure 16-7)

✦ **Mobango:** www.mobango.com (see Figure 16-8)

✦ **My-Symbian:** www.my-symbian.com (see Figure 16-9)

✦ **WildPalm:** www.wildpalm.co.uk (see Figure 16-10)

Figure 16-6: Nokia's Software Market offers games as well as other software.

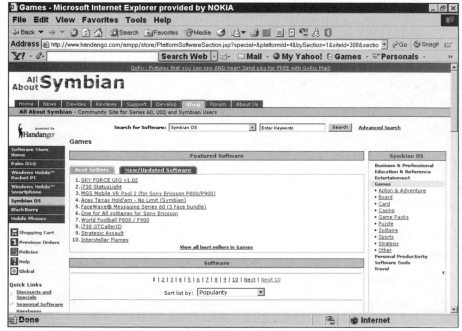

Figure 16-7: The All About Symbian page of Handango.com.

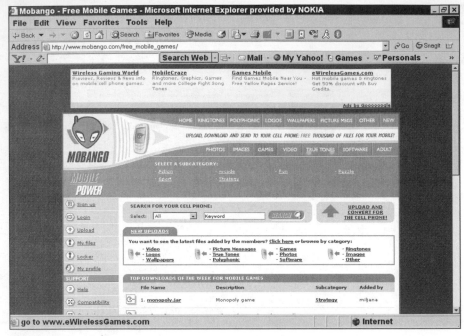

Figure 16-8: Mobango

Figure 16-9: My-Symbian

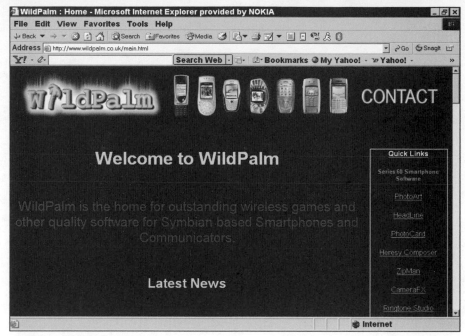

Figure 16-10: WildPalm

When you go hunting for phone games, browse these sites and take a look at what they offer. As with all the software mentioned in this book, try before you buy.

Buying and Using Accessories

The only accessories you could buy for your old cellphone were a leather case and maybe a wired headset (earphone and microphone combination that hangs rather ungracefully from the ear). Times have changed. Your new Series 60–based smartphone features a wide array of useful and cool accessories that can make you more productive and allow you to get more use out of your smartphone than you ever imagined.

In this chapter, you'll learn some shopping tips on the most popular Series 60–based smartphone accessories: car kits, headsets, and phone stands, as well as a few other wicked-cool phone enhancements. Unlike fashion accessories, phone accessories can empower you and allow you to use your phone in places and in ways you never imagined.

Technique 94: Buying and Using a Car Kit

How long is your commute? How much time do you spend in the car on weekends? If your answers are anywhere near "long" and "a lot" you should consider a car kit.

What exactly is a car kit? It's a combination speakerphone, microphone, and usually a dashboard mount to hold your phone securely. A car kit enables you to make and take calls while driving, without the need to hold your phone up against your ear.

Hands-free, eyes-free calling: It's not just a good idea . . .

It's the law! At least in some states. In New York State and in the pipeline in Massachusetts and other states, driving and using a cellphone without a car kit or headset is illegal. Why? A number of studies—and common sense—show a link between car accidents and driving while talking on a cellphone.

The solution? Invest in a car kit or a headset that will allow you to make and take calls while keeping your hands on the wheel and your eyes on the road. (See Technique 5 for tips on voice dialing.)

So-called aftermarket companies and phone makers themselves produce car kits for your smartphone. I recommend checking out both before finally settling on a particular car kit.

Here are the basic pros and cons you might encounter between phone-maker car kits and the aftermarket variety:

+ You might find that the manufacturer of your phone offers the car kit with the most features. Why? It's simple really: Your phone maker is most in touch with your phone's hardware and software and can therefore provide a car kit that takes the most advantage of your phone's capabilities. Aftermarket car kits are trying to support as many phones as possible in order to attract the largest group of customers. Thus, aftermarket car kits in most cases offer a least-common-denominator set of features.

+ Aftermarket car kits are commonly stocked in electronics stores and car-stereo installation centers. So, you can easily drop into one of these stores, pick out a car kit, and have it professionally installed on the same day. Buying a car kit from your phone's maker might require you to purchase it over the Web and then either install it yourself or ask if the installers in your local car-stereo store can professionally install it for you for a fee.

If, after investigating both options, you decide on an aftermarket car kit, ask about compatibility with your phone and check the return policy. Your goal is to make sure your phone will work properly with the car kit and that if, in fact, you encounter a problem, you can get your money back.

If you decide to go with a phone-maker's car kit, try to arrange for installation before you buy. You don't want to be stuck with a car kit that you can't use in your car.

Some phone makers like Nokia let you search for professional car-kit installers over their Web site. In this example, from NokiaUSA.com (www.nokiausa.com), you just pick your state from a list (see Figure 17-1), and it returns the names and contact information of installers in that state (see Figure 17-2).

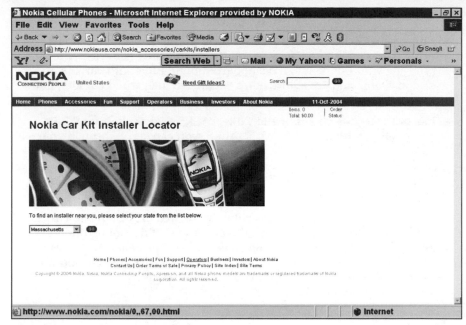

Figure 17-1: Pick your state.

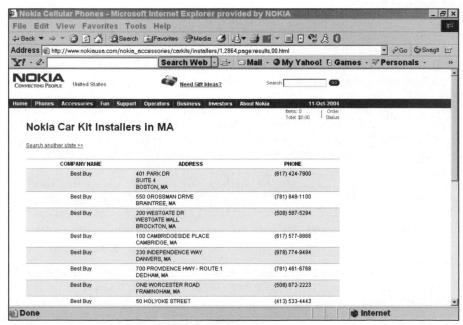

Figure 17-2: See the list of installers near you.

Here are some other tips for buying and using a car kit:

✦ **Consider a Bluetooth-based headset.** With Bluetooth, you can just bring your phone into your car and it's connected — no wires, no plugging your phone into a dashboard mount. Bluetooth car kits are also available and are a little more expensive than wired car kits, but the convenience is often worth the money.

✦ **Use a text-to-speech application like TALKS (see Technique 85) in conjunction with a car kit.** With TALKS, you can scroll through your contacts list while keeping your eyes on the road. The TALKS application reads your contact names to you, and when you come to the right one, you can just select it to dial without every needing to look at your phone's screen.

✦ **Even if you buy a car kit from your phone maker, be sure to check compatibility between your phone and the car kit.** Not all phones and car kits speak the same "language" or use the same connectors. Because of possible compatibility issues, make sure you ask about the return policy of any car kit you're considering.

Technique 95: Buying and Using a Headset

Even if you buy no other accessory, buy a headset for your smartphone, if one hasn't already been included in the box. What exactly is a headset? It's a combination of earphone and microphone that lets you talk to and hear from someone without having to hold the phone up to your face.

A headset is the ultimate hands-free accessory for your phone. Headsets range in price from $25 for a wired headset to $150 for a wireless version. These days, headsets offer a tremendous range of features including volume control, call answer and end, and even stereo sound and digital camera capability.

Personally, I think wireless Bluetooth headsets are the best. With a Bluetooth headset, you can actually leave your phone inside your belt carrying case, backpack, or briefcase and still make and take calls. By using voice tags or voice recognition technology you may not ever need to access your phone directly.

If your commute is long, but you can't justify a $200 car kit, consider a headset instead. They're cheaper than car kits but still allow you to operate your phone hands-free and eyes-free.

If your work entails a lot of walking but also involves your talking on your smartphone, consider a headset. Anyone who has held a phone up to his ear for 15 or 20 minutes knows how tiring long cellphone conversations can be on the arm.

Here are some additional tips for buying and using a headset:

✦ **As with car kits, you'll find both phone-maker and aftermarket headsets on the Web and in stores — look into both.** In this category of accessories, you don't have to find an installer like you do with car kits.

✦ **Always check for compatibility between your phone and the headset you're considering.** If you do purchase a headset that isn't made by your phone maker, make sure to ask about compatibility between your phone and the headset. Again, the connectors don't necessarily match. Most phone-maker Web sites let you pick your phone model and then display only the headsets (and other accessories) that are compatible with that phone.

✦ **Look for headsets that deliver features your phone doesn't.** Now you can purchase headphones that offer digital cameras, FM radios, stereo sound, and more. With the right headset you could add new functionality to your phone. Keep in mind, of course, that these types of headsets cost more than the standard type.

Technique 96: Buying and Using Stands and Other Accessories

Most older cellphones didn't have stands. You just plugged them into the wall outlet when you needed to recharge the batteries. Nowadays, phone stands still maintain that basic function but also add more.

So-called music stands let you play digital music or the FM radio (if your phone offers one) over speakers integrated into the stand. Some music stands let you plug additional audio gear — CD player or digital music player — into the back. Many let you use the additional speakers like a speakerphone for calling as well.

Connectivity stands make a cable connection (typically via USB) between your phone and your PC. *Remember:* You don't strictly need a cable connection because this book shows you how to connect your phone and PC using infrared and Bluetooth technology (see Chapter 9). But some people find connectivity stands easy and convenient (they automatically begin to synchronize contacts and calendar information with your PC as soon as you set your phone into the stand), and the stand still recharges your phone while it sits there.

Nokia even offers a video call stand for its 6630 Series 60 phone. It features a built-in camera for making streaming video calls.

Here are some tips for buying a phone stand:

+ **Make sure you really want one.** These smartphones don't strictly require a stand for recharging, making a PC connection, playing music, or anything else. There are techniques for doing all these things without a stand.

+ **For music stands, check the quality of the audio.** If possible, ask for a demonstration of the speakers using a plugged-in phone. *Remember:* These stands are meant to sit on a desk (either at work or at home), so you'll probably have access to other music-playing gear (even your PC or tabletop radio) that might even offer better sound quality. If, though, you find the music stands meets your needs, go for it.

+ **For connectivity stands, look for one that automatically, or with the touch of one button, synchronizes your phone and PC.** This convenience would be one of the main reasons to spend the money on a connectivity stand, so make sure it works seamlessly. And, again, you don't need to purchase a connectivity stand to synchronize files and data between your phone and a PC. But with a connectivity stand it's a little faster and easier.

Why keyboards are cool

One great accessory that I feel compelled to mention is a portable keyboard for your phone. Several phone makers now offer portable keyboards for their Series 60 phones, and I think they really make things easier.

For example, with a keyboard you can create longer text messages very quickly — much faster than with a standard phone keyboard (even using predictive text or some other speedup technique). You can also create and edit documents, e-mails, and notes. You can actually take notes at a meeting, as well.

And you don't lose anything in the way of portability. These portable keyboards fold and weigh only way a few ounces, so you can slip one right into a pocket, backpack, or briefcase and carry it with you wherever you go.

PDAs have sported portable keyboards for a few years now. I'm glad that smartphone makers are finally catching up by offering this important accessory.

Portable keyboards costs upwards of $100, but if you write a lot on the go, a portable keyboard will be well worth the money.

Trouble-shooting

Problems with your phone? No worries. We'll solve them together. I know general technical problems are frustrating. But when there's trouble with your smartphone — your connection to the world — it can drive you crazy.

Fortunately, your Series 60–based smartphone comes armed with utilities and special codes that will bring a sick phone back to full health in minutes.

In this chapter, I give you a closer look at the wireless personal networking technology called Bluetooth, including the most common problems with it and how to solve them. I clue you in to special codes that tell you things about your phone you didn't know. Technical-support people often use these codes in helping resolve issues. You'll also find out how to use your phone maker's online support information to solve problems yourself.

Technique 97: Troubleshooting a Low-Memory Warning

If you were to download all the applications described in this book and store them in your phone memory (rather than on a multimedia memory card as I suggest), you would definitely fill it up, and you would have a low-memory situation on your hands.

You'll know when you're in the midst of a low-memory situation. Your phone will present a message like `Memory low. Delete some data` or `Not enough memory to perform the operation. Delete some data first.`

I mentioned many of the following memory-saving tips in techniques throughout this book. But here is a systematic way to regain phone memory and bring your phone back to health:

✦ **Messages:** Clean out your SMS/MMS inbox. Use Technique 19 to save only the important multimedia elements you want to keep — and save those off to your multimedia memory card (MMC).

✦ **Images:** Go into your Gallery (Images on some Series 60 phones) and move images and video from your phone memory to your MMC using Technique 28. Or transfer the multimedia files to your PC using Technique 52.

✦ **Log:** Clean the phone's log files and, if you feel like you aren't getting good use from your log files (read Technique 29 for more information on how to use your log files), change the settings so that no new logs are created (see Technique 68).

✦ **Web browser:** Inside your phone's Web browser, clear the cache and delete all the saved Web pages. To clear cache on newer Series 60 models (Series 60 2nd Edition), choose Options ➪ Navigation Options. On older models (Series 60 1st Edition), choose Options ➪ Clear Cache.

✦ **Calendar:** Using Technique 64, clear out all your old calendar entries. Then reduce the amount of calendar information that gets downloaded from your PC during synchronization using Technique 89.

Great job! With all this cleaning you've probably saved anywhere from several hundred kilobytes up to several megabytes of data from your phone's memory. This cleaning should get your phone up and running and ready for action.

Technique 98: Gathering Information about Your Phone Using Secret Codes from Manufacturers

You may have heard of "Easter eggs" inside things like DVDs, PC games, and applications. By pressing a special combination of keys, you bring up an unexpected surprise that was inserted into the software by developers who created it.

Guess what? Your phone has "Easter eggs" of its own. No, they aren't cute graphics or a list of all the software developers that created the Series 60 user interface. These "Easter eggs" display information about your phone that is worth its weight in gold for troubleshooters and technical-support staffers.

Most phone user manuals don't list these codes at all. You might learn about them when you're interacting with a tech-support group over the phone or by e-mail to solve some phone problem. Or you may find out about them when you're trying to register your phone for a warranty or to obtain a software license for some special software.

I list them here so that you can become familiar with them and will quickly know what a tech-support person or phone-maker representative is asking for if you find yourself in that situation. This technique might end up making your phone call that much shorter — and that's always a good thing.

The secret codes all work basically the same way. They only work while the phone is in standby mode — waiting for you to make or take a call — not when you're viewing the main menu or inside any application.

Next, the codes all follow the same format. They all begin with the * key (bottom left on most phone keypads). The next character in the code is always the # key (bottom right on most keypads). Then you have a sequence of numbers. And, finally, the last key is also the # key.

The information that the code activates will be displayed in a pop-up window as soon as you press that last # key.

Still with me? Here is a list of the codes, the information each displays, and why each code is useful:

✦ **Code *#06#:** This shows your phone's serial number (also known as the IMEI number). When you obtain a license for certain software (like the TALKS application mentioned in Technique 85), you may need to provide your phone's serial number. You might also need your phone's serial number when you register for a warranty with the phone's manufacturer.

✦ **Code *#2820#:** This shows your phone's Bluetooth address. You might need to know your phone's Bluetooth address when troubleshooting a Bluetooth networking problem or when you set up a sophisticated office or home wireless network.

✦ **Code *#0000#:** This shows the version number of your phone's firmware. If you're discussing a phone repair with a manufacturer's tech-support staff, you might be asked to enter this key sequence and provide the firmware version running on your phone.

You may never need to use these secret codes, but if you ever do, you'll know how.

Technique 99: Finding Answers through Your Service Provider and Manufacturer's Online Technical Support

Each of the Series 60 phone makers and every service provider offers online support for its smartphones. If you're experiencing a problem, your service provider's and/or phone maker's Web sites should be the first places you go to for help. Phone makers and service providers all have a real interest in keeping their phone owners and users up and running. So, they posted some very helpful information online to help you get fast answers.

Of course, there is kind of a natural breakdown regarding the support they provide. Phone makers supply help about the phones themselves. Service providers generally post support information about the service or the phone network.

Here are the Series 60 phone-manufacturer Web sites:

+ **Nokia:** www.nokia.com

+ **Panasonic:** www.panasonic.com

+ **Samsung:** www.samsung.com

+ **Sendo:** www.sendo.com

+ **Siemens:** www.siemens-mobile.com

And here are some of the major service provider Web sites:

+ **AT&T Wireless:** www.attwireless.com

+ **Cingular:** www.cingular.com

+ **Orange:** www.orange.com

+ **T-Mobile:** www.t-mobile.com

+ **Verizon Wireless:** www.verizonwireless.com

+ **Vodafone:** www.vodafone.com

The last resort, of course, on both of those resources, is to phone their technical-support call centers. But save yourself some time and hassle by first searching the support sections of the sites.

Each phone-maker Web site features a support section. And within each is an updated FAQ section (or list of frequently asked questions). Search this list using the supplied categories or using keywords of your own. For example, if you're having trouble making an infrared connection between your phone and PC, search on "infrared" (see Figure 18-1, Figure 18-2, and Figure 18-3). Or, if you're having trouble sending a multimedia message, search on "messages" or "MMS."

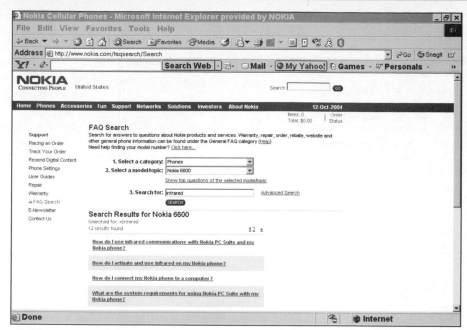

Figure 18-1: Searching the support section of Nokia's Web site for solutions to problems.

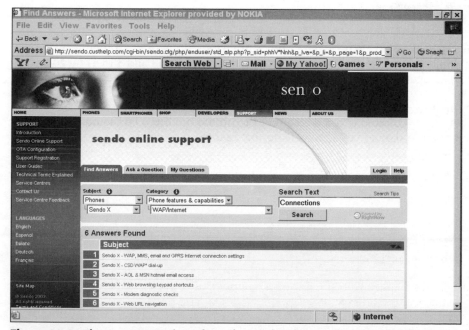

Figure 18-2: The support section of Sendo's Web site features a searchable knowledge base.

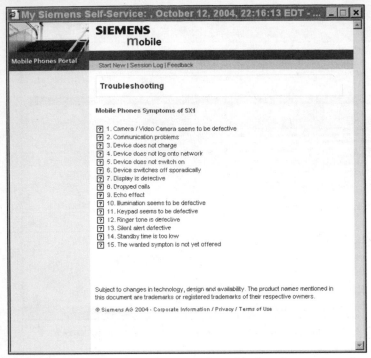

Figure 18-3: Siemens posts a special troubleshooting section of its site for the SX1.

Similarly, the service providers offer support sections or customer service on their sites. You can also browse through the FAQs (see Figure 18-4) on these sites or search the support databases using keywords (see Figure 18-5).

In the best case, you'll find exactly the information you need from these Web sites. But if you don't, you'll also find the phone numbers and e-mail addresses you need to reach live technical-support teams.

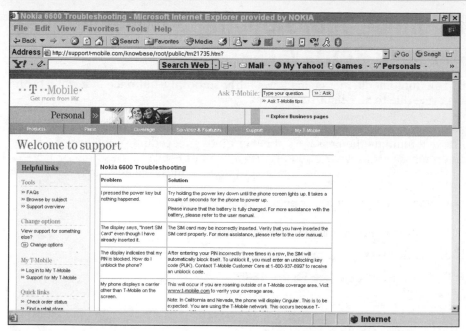

Figure 18-4: Browse T-Mobile's support section for information about your particular phone.

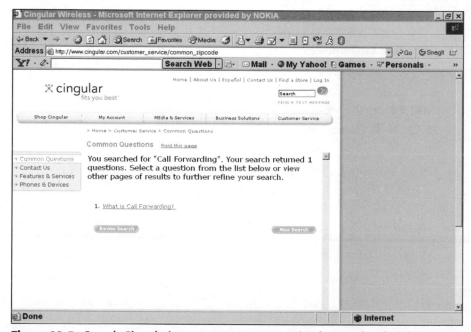

Figure 18-5: Search Cingular's customer care area using keywords related to your problem.

Technique 100: Stopping a Never-Ending Message Retrieval

All in all, multimedia messaging flows fairly smoothly, but once in a while problems arise due to size-limit restrictions.

One common symptom of such a problem is a phone trying to endlessly retrieve a multimedia message but never quite succeeding. You'll see periodic pop-up messages on your phone's screen like retrieving message or trying to retrieve message again, but no new message ever arrives.

Here's how to deal with never-ending message retrieval:

1. On the main menu, select **Messaging**.

2. Inside Messaging, press Options on the left menu key.

3. Choose Settings (see Figure 18-6).

Figure 18-6: Choose Settings.

4. Select Multimedia Message (MMS on some models), as shown in Figure 18-7.

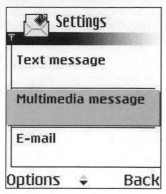

Figure 18-7: Select
Multimedia Message.

Tip   If you receive too many ads over your phone, change the multimedia message setting Receive Ads from Yes to No.

5. **Make sure the Access Point setting is set properly (not None, for example).**

   If in doubt, have a new MMS configuration message sent to your phone using Technique 2.

6. **To stop the never-ending retrieval, select On Receiving Message.**

7. **Choose Defer Retrieval (see Figure 18-8) if you want more time to try to diagnose the problem; if you want to temporarily shut off all multimedia message retrieval, choose Reject Message.**

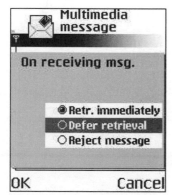

Figure 18-8: Defer Retrieval
will give you more time to
diagnose the problem.

When this problem happens to me I usually shut off all multimedia message retrieval for 24 hours or so and let the service provider address any problems in its network. Then, after I turn it back on, if the problem persists, I contact my service provider's technical support.

Technique 101: Troubleshooting a Bluetooth Connection

Bluetooth is one of the newest and coolest technologies that your phone delivers. But because it's new, it's unfamiliar to most people. And unfamiliarity breeds trouble that needs . . . well, troubleshooting!

Are you having difficulty making a Bluetooth connection with a PC or another phone? Here's a systematic way to uncover the problem. If you're trying to connect to a PC, make sure the PC's Bluetooth hardware is active and not hidden using the Windows Control Panel or the Bluetooth software that came with the Bluetooth adapter.

Here's how to check the Bluetooth settings on your phone:

1. **On the main menu, select Connectivity.**

2. **Inside Connectivity, select Bluetooth.**

3. **Make sure that the Bluetooth setting is On (see Figure 18-9).**

Figure 18-9: Make sure Bluetooth is On.

4. **Now double-check that My Phone's Visibility setting is set to Shown To All (see Figure 18-10).**

Figure 18-10: Make sure
My Phone's Visibility is
Shown To All.

Finally, make sure your phone and the other device are not more than 10
meters (30 feet) apart and that no thick walls are standing between them.

Index

Continued

Continued

Ringtones, Graphics, and Games for your Smartphone

ToneGuys.com employs proprietary Adaptive Content Server™ (ACS) technology to deliver optimized ringtones to your Smartphone.

Simply visit www.toneguys.com and run the free compatibility test Our ACS servers will automatically identify your exact cell phone model, optimize your ringtone, and deliver it to your phone in seconds.

FREE E-COUPON

Take advantage of ToneGuys free offer and select a free ringtone or graphic of your choice from our site. Our servers will optimize it for our cell phone and deliver it to your cell phone in seconds.

Visit www.toneguys.com, select a tone or graphic, and at checkout, simply use promotional code 'SMARTPHONE'.

ToneGuys is service mark of Unwired Appeal

Shot at 2:34

Shared at 2:35

Your life with instant replay. Shared instantly.

Shoot a basket. Shoot a movie. The Nokia video phone...a perfect way to save and share life's playful moments.
Feel connected.

Capture, store and send video
-video streaming
-digital camera
-email and high-speed Internet with EDGE
-sync calendar and contacts with PC
-Bluetooth® wireless

NOKIA 6620 video phone

NOKIA
CONNECTING PEOPLE

Save $20 off any order of $100 or more!
Visit www.nokia.com/us and enter
coupon code COOLCOUP at checkout.
Offer expires 12/31/06. See site for details.